STUDENT'S SOLUTIONS MANUAL

BEVERLY DRETZKE
University of Minnesota

ELEMENTARY STATISTICS USING EXCEL®
FIFTH EDITION

D1396668

Mario F. Triola
Dutchess Community College

PEARSON

Boston   Columbus   Indianapolis   New York   San Francisco   Upper Saddle River
Amsterdam   Cape Town   Dubai   London   Madrid   Milan   Munich   Paris   Montreal   Toronto
Delhi   Mexico City   São Paulo   Sydney   Hong Kong   Seoul   Singapore   Taipei   Tokyo

Copyright © 2014, 2010, 2007 Pearson Education, Inc.
Publishing as Pearson, 75 Arlington Street, Boston, MA 02116.

ISBN-13: 978-0-321-85167-3
ISBN-10: 0-321-85167-6

1 2 3 4 5 6 EBM 17 16 15 14 13

www.pearsonhighered.com

Contents

Chapter 1: Introduction to Statistics 1

Chapter 2: Summarizing and Graphing Data 7

Chapter 3: Statistics for Describing, Exploring, and Comparing Data........................ 19

Chapter 4: Probability 35

Chapter 5: Discrete Probability Distributions.. 47

Chapter 6: Normal Probability Distributions... 61

Chapter 7: Estimates and Sample Sizes... 85

Chapter 8: Hypothesis Testing.. 97

Chapter 9: Inferences from Two Samples 111

Chapter 10: Correlation and Regression.. 129

Chapter 11: Goodness-of-Fit and Contingency Tables 163

Chapter 12: Analysis of Variance... 173

Chapter 13: Nonparametric Statistics... 181

Chapter 14: Statistical Process Control 197

Chapter 1

Introduction to Statistics

Section 1-2, Basic Skills and Concepts

1. Statistical significance is indicated when methods of statistics are used to reach a conclusion that some treatment or finding is effective, but common sense might suggest that the treatment or finding does not make enough of a difference to justify its use or to be practical. Yes, it is possible for a study to have statistical significance but not a practical significance.

3. A voluntary response sample is a sample in which the subjects themselves decide whether to be included in the study. A voluntary response sample is generally not suitable for a statistical study because the sample may have a bias resulting from participation by those with a special interest in the topic being studied.

5. There does appear to be a potential to create a bias.

7. There does not appear to be a potential to create a bias.

9. The sample is a voluntary response sample and is therefore flawed.

11. The sampling method appears to be sound.

13. Because there is a 30% chance of getting such results with a diet that has no effect, it does not appear to have statistical significance, but the average loss of 45 pounds does appear to have practical significance.

15. Because there is a 23% chance of getting such results with a program that has no effect, the program does not appear to have statistical significance. Because the success rate of 23% is not much better than the 20% rate that is typically expected with random guessing, the program does not appear to have practical significance.

17. The male and female pulse rates in the same column are not matched in any meaningful way. It does not make sense to use the difference between any of the pulse rates that are in the same column.

19. The data can be used to address the issue of whether males and females have pulse rates with the same average (mean) value.

21. Yes, each IQ score is matched with the brain volume in the same column, because they are measurements obtained from the same person. It does not make sense to use the difference between each IQ score and the brain volume in the same column, because IQ scores and brain volumes use different units of measurement. For example, it would make no sense to find the difference between an IQ score of 87 and a brain volume of 1035 cm^3.

23. Given that the researchers do not appear to benefit from the results, they are professionals at prestigious institutions, and funding is from a U.S. government agency, the source of the data appears to be unbiased.

25. It is questionable that the sponsor is the Idaho Potato Commission and the favorite vegetable is potatoes.

27. The correlation, or association, between two variables does not mean that one of the variables is the cause of the other. Correlation does not imply causation.

29. a. The number of people is $(0.39)(1018) = 397.02$.

    b. No. Because the result is a count of people among 1018 who were surveyed, the result must be a whole number.

    c. The actual number is 397 people.

    d. The percentage is $\dfrac{255}{1018} = 0.25049 = 25.049\%$.

31. a. The number of adults is $(0.14)(2302) = 322.28$

    b. No. Because the result is a count of adults among 2302 who were surveyed, the result must be a whole number.

    c. The actual number is 322 adults.

d.   The percentage is $\dfrac{46}{2302} = 0.01998 = 1.998\%$.

33. Because a reduction of 100% would eliminate all of the size, it is not possible to reduce the size by 100% or more.

35. If foreign investment fell by 100% it would be totally eliminated, so it is not possible for it to fall by more than 100%.

Section 1-2, Beyond the Basics

37. Without our knowing anything about the number of ATVs in use, or the number of ATV drivers, or the amount of ATV usage, the number of 740 fatal accidents has no context. Some information should be given so that the reader can understand the rate of ATV fatalities.

39. The wording of the question is biased and tends to encourage negative response. The sample size of 20 is too small. Survey respondents are self-selected instead of being selected by the newspaper. If 20 readers respond, the percentages should be multiples of 5, so 87% and 13% are not possible results.

Section 1-3, Basic Skills and Concepts

1. A parameter is a numerical measurement describing some characteristic of a population, whereas a statistic is a numerical measurement describing some characteristic of a sample.

3. Parts (a) and (c) describe discrete data.

5. Statistic

7. Parameter

9. Parameter

11. Statistic

13. Continuous

15. Discrete

17. Discrete

19. Continuous

21. Nominal

23. Interval

25. Ratio

27. Ordinal

29. The numbers are not counts or measures of anything, so they are at the nominal level of measurement, and it makes no sense to compute the average (mean) of them.

31. The numbers are used as substitutes for the categories of low, medium, and high, so the numbers are at the ordinal level of measurement. It does not make sense to compute the average (mean) of such numbers.

Section 1-3, Beyond the Basics

33. a.   Continuous, because the number of possible values is infinite and not countable.

b.   Discrete, because the number of possible values is finite.

c.   Discrete, because the number of possible values is finite.

d.   Discrete, because the number of possible values is finite and countable.

35. With no natural starting point, temperatures are at the interval level of measurement so ratios such as "twice" are meaningless.

Section 1-4, Basic Skills and Concepts

1. No. Not every sample of the same size has the same chance of being selected. For example, the sample with the first two names has no chance of being selected. A simple random sample of n items is selected in such a way that every sample of the same size has the same chance of being selected.

3. The population consists of the adult friends on the list. The simple random sample is selected from the population of adult friends on the list, so the results are not likely to be representative of the much larger general population of adults in the United States.

5. Because the research participants are subjected to anger and confrontation, they are given a form of treatment, so this is an experiment, not an observational study.

7. This is an observational study because the therapists were not given any treatment. Their responses were observed.

9. Cluster

11. Random

13. Convenience

15. Systematic

17. Random

19. Convenience

21. The sample is not a simple random sample. Because every 1000th pill is selected, some samples have no chance of being selected. For example, a sample consisting of two consecutive pills has no chance of being selected, and this violates the requirement of a simple random sample.

23. The sample is a simple random sample. Every sample of size 500 has the same chance of being selected.

25. The sample is not a simple random sample. Not every sample has the same chance of being selected. For example, a sample that includes people who do not appear to be approachable has no chance of being selected.

Section 1-4, Beyond the Basics

27. Prospective study

29. Cross-sectional study

31. Matched pairs design

33. Completely randomized design

35. Blinding is a method whereby a subject (or person who evaluates results) in an experiment does not know whether the subject is treated with the DNA vaccine or the adenoviral vector vaccine. It is important to use blinding so that results are not somehow distorted by knowledge of the particular treatment used.

Section 1-5, Basic Skills and Concepts

1. A spreadsheet is a collection of data organized in an array of cells arranged in rows and columns, and it is used to summarize, analyze, and perform calculations with the data.

3. An Excel file that contains worksheets

5. The top half of the worksheet is deleted, then restored.

7. The top half of the worksheet is deleted, then restored.

9. The described worksheet is printed.

11. The described worksheet is printed.

Section 1-5, Beyond the Basics

13. If using Excel 2013, 2010, or 1007, click the File tab; then select Save As. If using Excel 2007, click on the Office Button; then Select Save As. Enter the new file name and location for the saved file.

Chapter Quick Quiz

1. No. The numbers do not measure anything.

2. Nominal

3. Continuous

4. Quantitative data

5. Ratio

6. False

7. No

8. Statistic

9. Observational study

10. False

Review Exercises

1. a. Discrete

    b. Ratio

    c. Stratified

    d. Cluster

    e. The mailed responses would be a voluntary response sample, so those with strong opinions are more likely to respond. It is very possible that the results do not reflect the true opinions of the population of all customers.

2. The survey was sponsored by the American Laser Centers, and 24% said that their favorite body part is the face, which happens to be a body part often chosen for some type of laser treatment. The source is therefore questionable.

3. The sample is a voluntary response sample, so the results are questionable.

4. a. It uses a voluntary response sample, and those with special interests are more likely to respond, so it is very possible that the sample is not representative of the population.

    b. Because the statement refers to 72% of all Americans, it is a parameter (but it is probably based on a 72% rate from the sample, and the sample percentage is a statistic).

    c. Observational study

5. a. If they have no fat at all, they have 100% less than any other amount with fat, so the 125% figure cannot be correct.

    b. The exact number is (0.58)(1182) = 685.56. The actual number is 686.

    c. $\dfrac{331}{1182} = 0.28003 = 28.003\%$

6. The Gallop poll used randomly selected respondents, but the AOL poll used a voluntary response sample. Respondents in the AOL poll are more likely to participate if they have strong feelings about the candidates, and this group is not necessarily representative of the population. The results from the Gallop poll are more likely to reflect the true opinions of American voters.

7. Because there is only a 4% chance of getting the results by chance, the method appears to have statistical significance. The results of 112 girls in 200 births is above the approximately 50% rate expected by chance, but

it does not appear to be high enough to have practical significance. Not many couples would bother with a procedure that raises the likelihood of a girls from 50% to 56%.

8. a. Random

    b. Stratified

    c. Nominal

    d. Statistic, because it is based on a sample.

    e. The mailed responses would be a voluntary response sample. Those with strong opinions about the topic would be more likely to respond, so it is very possible that the results would not reflect the true opinions of the population of all adults.

9. a. Systematic

    b. Random

    c. Cluster

    d. Stratified

    e. Convenience

    f. No, although this is a subjective judgment.

10. a. $(0.52)(1500) = 780$ adults

    b. $\dfrac{345}{1500} = 0.23 = 23\%$

    c. Men: $\dfrac{727}{1500} = 0.485 = 48.5\%$

        Women: $\dfrac{773}{1500} = 0.515 = 51.5\%$

Cumulative Review Exercises

1. You will use Excel's AVERAGE function to solve the problem. Key in the formula =**AVERAGE(A2:A49)** in an Excel worksheet and press [**Enter**]. The mean is equal to 11. Because the flight numbers are not measures or counts of anything, the result does not have meaning.

2. You will use Excel's AVERAGE function to solve the problem. Key in the formula) =**AVERAGE(D2:D21** in an Excel worksheet and press [**Enter**]. The mean is equal to 101, and it is reasonably close to the population mean of 100.

3. You will use Excel's STANDARDIZE function to solve the problem. Key in the formula =**STANDARDIZE(247,176,6)** in an Excel worksheet and press [**Enter**]. The standardized score is equal to 11.83 which is an unusually high value.

4. Key in the formula =**(175-172)/(29/SQRT(20))** in an Excel worksheet and press [**Enter**]. The result is 0.46.

5. Key in the formula =**1.96^2*0.25/0.03^2** in an Excel worksheet and press [**Enter**]. The result is 1067.

6. Key in the formula =**(88-88.57)^2/88.57** in an Excel worksheet and press [**Enter**]. The result is 0.0037.

7. Key in the formula =**((96-100)^2+(106-100)^2+(98-100)^2)/(3-1)** in an Excel worksheet and press [**Enter**]. The result is 28.

8. Key in the formula =**SQRT(((96-100)^2+(106-100)^2+(98-100)^2)/(3-1))** in an Excel worksheet and press [**Enter**]. The result is 5.3.

9. Key in the formula =**0.6^14** in an Excel worksheet and press [**Enter**]. The result is 0.000783642.

10. Key in the formula =8^12 in an Excel worksheet and press [**Enter**]. The result is 68719476736.

11. Key in the formula =**7^14** in an Excel worksheet and press [**Enter**]. The result in scientific notation is 6.78223E+11. The result, not in scientific notation, is 678223072849. Hint: Change the format from General to Number to view the result as an ordinary number, not in scientific notation.

12. Key in the formula =**0.3^10** in an Excel worksheet and press [**Enter**]. The result in scientific notation is 5.9049E-06. The result, not in scientific notation, is 0.00000590493. Hint: Change the format from General to Number and set the number of decimal places at 10 to view the result as an ordinary number, not in scientific notation.

Chapter 2

Summarizing and Graphing Data

Section 2-2, Basic Skills and Concepts

1. No. For each class, the frequency tells us how many values fall within the given range of values, but there is no way to determine the exact IQ scores represented in the class.

3. No. The sum of the percentages is 199% not 100%, so each respondent could answer "yes" to more than one category. The table does not show the distribution of a data set among all of several different categories. Instead, it shows responses to five separate questions.

5. Class width: 10.

   Class midpoints: 24.5, 34.5, 44.5, 54.5, 64.5, 74.5, 84.5.

   Class boundaries: 19.5, 29.5, 39.5, 49.5, 59.5, 69.5, 79.5, 89.5.

7. Class width: 10.

   Class midpoints: 54.5, 64.5, 74.5, 84.5, 94.5, 104.5, 114.5, 124.5.

   Class boundaries: 49.5, 59.5, 69.5, 79.5, 89.5, 99.5, 109.5, 119.5, 129.5.

9. Class width: 2.

   Class midpoints: 3.95, 5.95, 7.95, 9.95, 11.95.

   Class boundaries: 2.95, 4.95, 6.95, 8.95, 10.95, 12.95.

11. No. The frequencies do not satisfy the requirement of being roughly symmetric about the maximum frequency of 34.

13. 18, 7, 4.

15. On average, the actresses appear to be younger than the actors.

Age When Oscar Was Won	Relative Frequency (Actresses)	Relative Frequency (Actors)
20 – 29	32.9%	1.2%
30 – 39	41.5%	31.7%
40 – 49	15.9%	42.7%
50 – 59	2.4%	15.9%
60 – 69	4.9%	7.3%
70 – 79	1.2%	1.2%
80 – 89	1.2%	0.0%

17. The cumulative frequency table is

Age (years) of Best Actress When Oscar Was Won	Cumulative Frequency
Less than 30	27
Less than 40	61
Less than 50	74
Less than 60	76
Less than 70	80
Less than 80	81
Less than 90	82

19. Enter the label **Last Digit** in cell A1 of an Excel worksheet followed by the last digit data. Select the **XLSTAT** add-in. Select **Describing data**. Select **Histograms**. Enter the data range including the label (**A1:A38**). Select **Discrete**. Select **Sample labels**. Click the **Options** tab. Select **Number** of intervals and enter **10**. Click **OK**. The first three columns of the XLSTAT output are displayed below. The values in the lower bound columns are the last digits.

Because there are disproportionately more 0s and 5s, it appears that the heights were reported instead of measured. Consequently, it is likely that the results are not very accurate.

Lower bound	Upper bound	Frequency
0	1	9
1	2	2
2	3	1
3	4	3
4	5	1
5	6	15
6	7	2
7	8	0
8	9	3
9	10	1

21. Open the **MBODY** data file on your data disk. Enter the label **Lower** in cell P1 followed by the lower class limits of 40, 50, 60, 70, 80, 90, and 100, as displayed on the right. Select the **XLSTAT** add-in. Select **Describing data**. Select **Histograms**. Enter the range of the PULSE data including the label (**B1:B41**). Select **Discrete**. Select **Sample labels**. Click the **Options** tab. Select **User defined** and enter range of the user defined lower class limits (**P1:P8**). Click **OK**. The first three columns of the XLSTAT output are displayed below. Yes, the distribution appears to be a normal distribution.

P
Lower
40.0
50.0
60.0
70.0
80.0
90.0
100.0

Lower bound	Upper bound	Frequency
40	50	1
50	60	7
60	70	17
70	80	9
80	90	5
90	100	1

23. Open the **QUAKE** data file on your data disk. Enter the label **Lower** in cell D1 followed by the lower class limits of 0.0, 0.5, 1.0, 1.5, 2.0, 2.5, and 3.0 as displayed below on the left. Select the **XLSTAT** add-in. Select **Describing data**. Select **Histograms**. Enter the range of the MAG data including the label (**A1:A51**). Select **Continuous**. Select **Sample labels**. Click the **Options** tab. Select **User defined** and enter range of the user defined lower class limits (**D1:D8**). Click **OK**. The first three columns of the XLSTAT output are displayed below on the right. No, the distribution does not appear to be a normal distribution.

D
Lower
0.0
0.5
1.0
1.5
2.0
2.5
3.0

Lower bound	Upper bound	Frequency
0	0.5	5
0.5	1	15
1	1.5	19
1.5	2	7
2	2.5	2
2.5	3	2

25. Open the **MBODY** data file on your data disk. Enter the label **Lower** in cell P1 followed by the lower class limits of 4.0, 4.4, 4.8, 5.2, 5.6, and 6.0 as displayed on the right. Select the **XLSTAT** add-in. Select **Describing data**. Select **Histograms**. Enter the range of the RED data including the label (**H1:H41**). Select **Continuous**. Select **Sample labels**. Click the **Options** tab. Select **User defined** and enter range of the user defined lower class limits (**P1:P7**). Click **OK**. The first three columns of the XLSTAT output are displayed. Yes, the distribution appears to be roughly a normal distribution.

P
Lower
4.0
4.4
4.8
5.2
5.6
6.0

Lower bound	Upper bound	Frequency
4	4.4	2
4.4	4.8	7
4.8	5.2	15
5.2	5.6	13
5.6	6	3

27. Open the **FLIGHTS** data file on your data disk. Enter the label **Lower** in cell H1 followed by the lower class limits as displayed below. Select the **XLSTAT** add-in. Select **Describing data**. Select **Histograms**. Enter the range of the Arr Delay data including the label (**F1:H49**). Select **Discrete**. Select **Sample labels**. Click the **Options** tab. Select **User defined** and enter range of the user defined lower class limits (**H1:H8**). Click **OK**. The first three columns of the XLSTAT output are displayed below on the right. Yes. Among the 48 flights, 36 arrived on time or early, and 45 of the flights arrived no more than 30 minutes late.

H
Lower
-60
-30
0
30
60
90
120

Lower bound	Upper bound	Frequency
-60	-30	11
-30	0	25
0	30	9
30	60	1
60	90	0
90	120	2

29. Enter the categories and frequencies in an Excel worksheet. Then use formulas to compute the relative frequencies. The completed table is displayed on the right.

Category	Frequency	Relative Frequency
Male Survivors	361	16.2%
Males Who Died	1395	62.8%
Female Survivors	345	15.5%
Females Who Died	122	5.5%
Total	2223	100.0%

31. Pilot error is the most serious threat to aviation safety. Better training and stricter pilot requirements can improve aviation safety.

Cause	Relative Frequency
Pilot Error	50.5%
Other Human Error	6.1%
Weather	12.1%
Mechanical	22.2%
Sabotage	9.1%

Section 2-2, Beyond the Basics

33. Open the **CANS** data file on your data disk. Enter the label **Lower** in cell D1 followed by the lower class limits displayed on the right. Select the **XLSTAT** add-in. Select **Describing data**. Select **Histograms**. Enter the range of the CANS111 data including the label (**B1:B176**). Select **Continuous**. Select **Sample labels**. Click the **Options** tab. Select **User defined** and enter range of the user defined lower class limits (**D1:D18**). Click **OK**. The first three columns of the XLSTAT output including the outlier are displayed below on the left. Next construct the distribution excluding the outlier. Make the necessary adjustments to the input ranges. The first three columns of the XLSTAT excluding the outlier are displayed below on the right. An outlier can dramatically affect the frequency table.

D
Lower
200
220
240
260
280
300
320
340
360
380
400
420
440
460
480
500
520

Lower bound	Upper bound	Frequency
200	220	6
220	240	5
240	260	12
260	280	36
280	300	87
300	320	28
320	340	0
340	360	0
360	380	0
380	400	0
400	420	0
420	440	0
440	460	0
460	480	0
480	500	0
500	520	1

Lower bound	Upper bound	Frequency
200	220	6
220	240	5
240	260	12
260	280	36
280	300	87
300	320	28

Section 2-3, Basic Skills and Concepts

1. It is easier to see the distribution of the data by examining the graph of the histogram than by the numbers in the frequency distribution.

3. With a data set that is so small, the true nature of the distribution cannot be seen with a histogram. The data set has an outlier of 1 minute. That duration time corresponds to the last flight, which ended in an explosion that killed seven crew members.

5. Identifying the exact value is not easy, but answers not too far from 200 are good answers.

7. The tallest person is about 108 inches, or about 9 feet tall. That tallest height is depicted in the bar that is farthest to the right in the histogram. That height is an outlier because it is very far from all of the other heights. The height of 9 feet must be an error, because the height of the tallest human ever recorded was 8 feet 11 inches.

9. Enter the label **Last Digit** in cell A1 of an Excel worksheet followed by the last digit data. Select the **XLSTAT** add-in. Select **Describing data**. Select **Histograms**. Enter the data range including the label (**A1:A38**). Select **Discrete**. Select **Sample labels**. Click the **Options** tab. Select **Number** of intervals and enter **10**. Click the **Charts** tab. Select **Histogram** and **Bars**. Select **Frequency** for the ordinate of the histogram. Click **OK**.

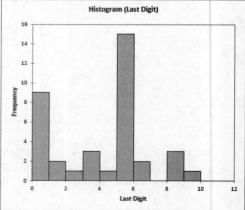

The digits 0 and 5 seem to occur much more than the other digits, so it appears that the heights were reported and not actually measured. This suggests that the results might not be very accurate.

11. Open the **MBODY** data file on your data disk.
Enter the label **Lower** in cell P1 followed by the
lower class limits as displayed on the right. Select
the **XLSTAT** add-in. Select **Describing data**.
Select **Histograms**. Enter the range of the PULSE
data including the label (**B1:B41**). Select **Discrete**.
Select **Sample labels**. Click the **Options** tab.
Select **User defined** and enter range of the user
defined lower class limits (**P1:P8**). Click the
Charts tab. Select **Histogram** and **Bars**. Select **Frequency** for
the ordinate of the histogram. Click **OK**. The histogram does
appear to depict a normal distribution. The frequencies
increase to a maximum and then tend to decrease, and the
histogram is symmetric with the left half being roughly a
mirror image of the right half.

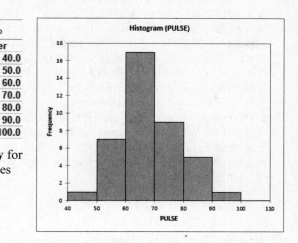

13. Open the **QUAKE** data file on your data disk. Enter the label **Lower** in cell D1 followed by the lower class
limits displayed on the right. Select the **XLSTAT** add-in. Select **Describing data**. Select **Histograms**. Enter the
range of the MAG data including the label (**A1:A51**). Select **Continuous**. Select **Sample labels**. Click the
Options tab. Select **User defined** and enter range of the user defined lower class limits (**D1: D8**). Click the
Charts tab. Select **Histogram** and **Bars**. Select **Frequency** for the ordinate of the histogram. Click **OK**. The
histogram is displayed below. The histogram appears to roughly approximate a normal distribution. The
frequencies increase to a maximum and then tend to decrease, and the histogram is symmetric with the left half
being roughly a mirror image of the right half.

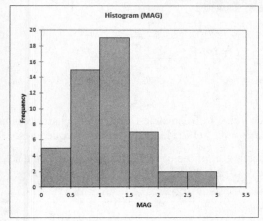

15. Open the **MBODY** data file on your data disk. Enter the
label **Lower** in cell P1 followed by the lower class limits
of 4.0, 4.4, 4.8, 5.2, 5.6, and 6.0 as displayed on the right.
Select the **XLSTAT** add-in. Select **Describing data**.
Select **Histograms**. Enter the range of the RED data
including the label (**H1:H41**). Select **Continuous**. Select
Sample labels. Click the **Options** tab. Select **User
defined** and enter range of the user defined lower class
limits (**P1:P7**). Click the **Charts** tab. Select **Histogram** and **Bars**.
Select **Frequency** for the ordinate of the histogram. Click **OK**. The
histogram appears to roughly approximate a normal distribution. The
frequencies increase to a maximum and then tend to decrease, and the
histogram is symmetric with the left half being roughly a mirror image
of the right half.

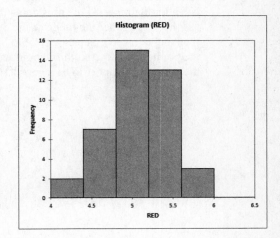

17. Open the **FLIGHTS** data file on your data disk.
     Enter the label **Lower** in cell H1 followed by the
     lower class limits as displayed on the right. Select
     the **XLSTAT** add-in. Select **Describing data**.
     Select **Histograms**. Enter the range of the Arr
     Delay data including the label (**F1:H49**). Select
     Discrete. Select **Sample labels**. Click the **Options**
     tab. Select **User defined** and enter range of the
     user defined lower class limits (**H1:H8**). Click the
     Charts tab. Select **Histogram** and **Bars**. Select **Frequency** for
     the ordinate of the histogram. Click **OK**. The two leftmost bars
     depict flights that arrived early, and the other bars to the right
     depict flights that arrived late.

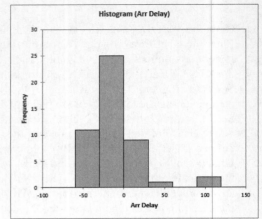

Section 2-3, Beyond the Basics

19. The ages of actresses are lower than those of actors.

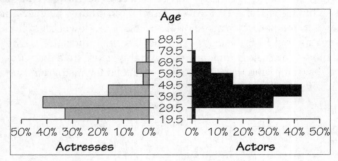

Section 2-4, Basic Skills and Concepts

1.   In a Pareto chart, the bars are arranged in descending order according to frequencies. The Pareto chart helps us
     understand data by drawing attention to the more important categories, which have the highest frequencies.

3.   The data set is too small for a graph to reveal important characteristics of the data. With such a small data set, it
     would be better to simply list the data or place them in a table.

5.   Open the **POTUS** data file on your data disk. Select the
     XLSTAT add-in. Select **Visualizing data**. Select **Scatter plots**.
     X: Enter the range of the presidents' heights including the label
     (E1:E39). Y: Enter the range of the opponents' heights including
     the label (F1:F39). Select **Variable labels**. Click **OK**. Because
     the points are scattered throughout with no obvious pattern,
     there does not appear to be a correlation.

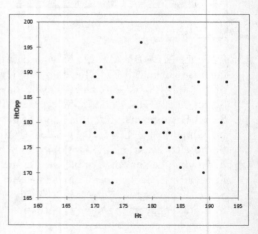

7. Open the **BEARS** data file on your data disk. Select the
    XLSTAT add-in. Select **Visualizing data**. Select **Scatter plots**.
    X: Enter the range of the chest size data including the label
    (H1:H55). Y: Enter the range of the weight data including the
    label (I1:I55). Select **Variable labels**. Click **OK**. Yes. There is
    a very distinct pattern showing that bears with larger chest sizes
    tend to weigh more.

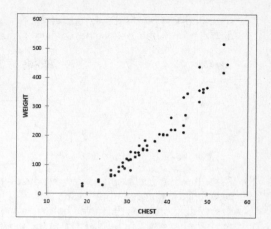

9. Enter the label **Day of Release** in cell A1 of an Excel worksheet followed by the numbers 1 through 14. Enter
    the label **Gross (millions of dollars)** in cell B1 followed by the amounts given in Exercise 9. Select the
    XLSTAT add-in. Select **Visualizing data**. Select **Scatter plots**. X: Enter the range of the Day of Release data
    including the label (A1:A15). Y: Enter the range of the Gross (millions of dollars) data including the label
    (B1:B15). Select **Variable labels**. Click **OK**. The scatter plot is displayed below. The first amount is highest for
    the opening day, when many Harry Potter fans are most eager to see the movie; the third and fourth values are
    from the first Friday and the first Saturday, which are the popular weekend days when movie attendance tends
    to spike.

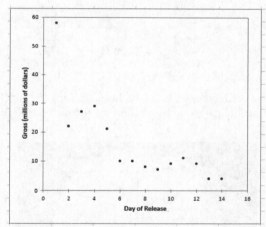

11. Yes, because the configuration of the points is roughly a bell shape, the volumes appear to be from a normally
    distributed population. The volume of 11.8 oz. appears to be an outlier.

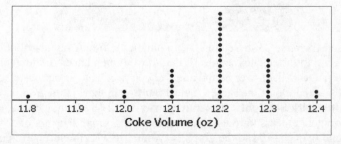

13. Open the **CRASH** data file on your data disk. Select the **XLSTAT** add-in. Select **Describing data**. Select
    Descriptive statistics. Enter the range of the PLVS data including the label (H1:H22). Select **Sample labels**.
    Click the **Options** tab. Select **Charts**. Click the **Charts (1)** tab. Select **Stem-and-leaf plots**. Click **OK**. The

stem-and-leaf plot is displayed below. No. The distribution is not dramatically far from being a normal distribution with a bell shape, so there is not strong evidence against a normal distribution

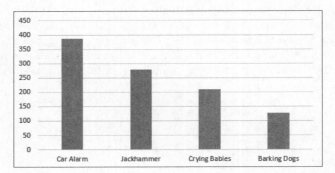

15. Enter the four frustrating sound responses in column A of an Excel worksheet (A1:A4). Enter the frequencies associated with each of the responses in column B (B1:B4). Sort the two columns by frequency arranged in descending order as displayed below on the left. Click and drag over the range A1:B4 to select these cells for the pareto chart. Click **INSERT** at the top of the screen. Select Column Chart in the Charts group by clicking on the column chart figure. Select the leftmost figure in the 2-D Column. The pareto chart is displayed below on the right.

A	B
Response	**Frequency**
Car Alarm	388
Jackhammer	279
Crying Babies	209
Barking Dogs	128

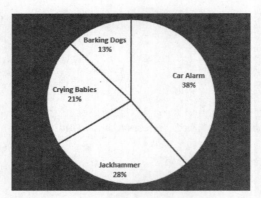

17. Click and drag over the Exercise 15 data range to select these cells for the pie chart. Click **INSERT** at the top of the screen. Select Pie Chart in the Charts group by clicking on the pie chart figure. Select the leftmost figure in the 2-D Pie row.

19. Follow the instructions in Exercise 23 in Section 2-2 to obtain the frequency distribution output. Cut and paste the lower bound values and the frequency values so that they are in adjacent columns of an Excel worksheet. Click and drag over the range of the lower bound and frequency cells to select these cells for the frequency polygon. Click **INSERT** at the top of the screen. Select **Scatter** in the Charts group. Select the rightmost figure in the second row, **Scatter with Straight Lines and Markers**. The frequency polygon is displayed on the next page. The frequency polygon appears to roughly approximate a normal distribution. The frequencies increase to a maximum and then tend to decease, and the graph is symmetric with the left half being roughly a mirror image of the right half.

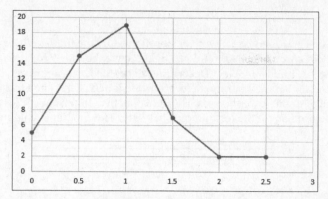

21. The vertical scale does not start at 0, so the difference is exaggerated. The graphs make it appear that Obama got about twice as many votes as McCain, but Obama actually got about 69 million votes compared to 60 million to McCain.

23. China's oil consumption is 2.7 times (or roughly 3 times) that of the United States, but by using a larger barrel that is three times as wide and three times as tall (and also three times as deep) as the smaller barrel, the illustration has made it appear that the larger barrel has a volume that is 27 times that of the smaller barrel. The actual ratio of US consumption to China's consumption is roughly 3 to 1, but the illustration makes it appear to be 27 to 1.

Section 2-4, Beyond the Basics

25. The ages of actresses are lower than those of actors.

```
                            Actresses         Actors
        9999999888877776666 55554421 | 2 | 9
99888877655555554443 3333222111000 | 3 | 00122244455666777788888999
            9655322111110 | 4 | 000111111222222333344455 55677788999
                       40 | 5 | 0112222346677
                     3110 | 6 | 000222
                        4 | 7 | 6
                        0 | 8 |
```

Chapter Quick Quiz

1. The class width is 1.00

2. The class boundaries are -0.005 and 0.995

3. No

4. 61 min., 62 min., 62 min., 62 min., 62 min., 67 min., and 69 min.

5. No

6. Bar graph

7. Scatterplot

8. Pareto Chart

9. The distribution of the data set

10. The bars of the histogram start relatively low, increase to a maximum value and then decrease. Also, the histogram is symmetric with the left half being roughly a mirror image of the right half.

Review Exercises

1. Enter the label **cm^3** in cell A1 of an Excel worksheet followed by the brain volume data given in Review Exercise 1. Enter the label **Lower** in cell B1followed by the lower bounds 900, 1000, 1100, 1200, 1300, 1400, and 1500. Select the **XLSTAT** add-in. Select **Describing data**. Select **Histograms**. Enter the data range including the label (**A1:A21**). Select **Discrete**. Select **Sample labels**. Click the **Options** tab. Select **User defined** and enter the range of the lower bounds, B1:B6. Click the **Charts** tab. Select **Histograms** and **Bars**. Select **Frequency** for the ordinate of the histogram. Click **OK**. The first three columns of the XLSTAT output are displayed below.

Lower bound	Upper bound	Frequency
900	1000	1
1000	1200	14
1200	1300	3
1300	1400	1
1400	1500	1

2. The histogram was included in the Review Exercise 1 output. The histogram is displayed below. No, the distribution does not appear to be normal because the graph is not symmetric.

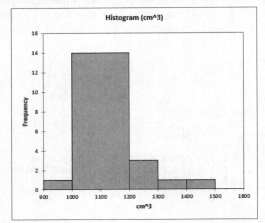

3. Although there are differences among the frequencies of the digits, the differences are not too extreme given the relatively small sample size, so the lottery appears to be fair.

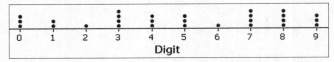

4. The sample size is not large enough to reveal the true nature of the distribution of IQ scores for the population from which the sample is obtained.

    8 | 7 7 9

    9 | 6 6

    10 | 1 3 3

5. A time-series graph is best. It suggests that the amounts of carbon monoxide emissions in the United States are increasing. To construct a time-series graph, begin by entering the label **Year** in cell A1 of an Excel worksheet followed by the numbers 1 through 10. Enter the label **Carbon Monoxide Emissions** followed by the data given in Review Exercise 5. Click and drag over the data range including labels (A1:B11). Click **INSERT** at the top of the screen. Select **Scatter** in the **Charts** group. Select Scatter with Straight Lines and Markers. The time-series graph is displayed at the top of the next page.

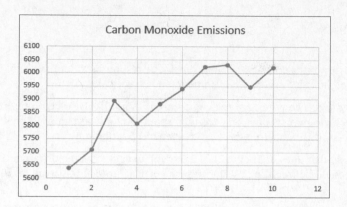

6. A scatterplot is best. The scatterplot does not suggest that there is a relationship. To construct a scatterplot, continue using the Excel worksheet that you constructed for Review Exercise 5. Add a column with the nitrious oxide emissions data as shown below on the left. Click and drag over the range of the carbon monoxide emissions and nitrious oxide emissions to select these data for the graph. Click **INSERT** at the top of the screen. Select Scatter in the Charts group. If you would like, add horizontal and vertical axis titles.

Year	Carbon Monoxide Emissions	Nitrious Oxide Emissions
1	5638	351
2	5708	349
3	5893	345
4	5807	339
5	5881	335
6	5939	335
7	6024	362
8	6032	371
9	5946	376
10	6022	384

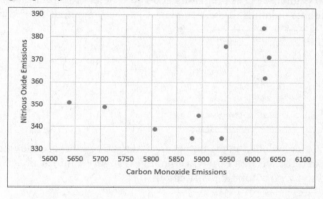

7. A Pareto chart is best. To construct a Pareto chart, enter the data in an Excel worksheet and sort the data by sales in descending order. Click and drag over the category and sales data to select the cells. Click **INSERT** at the top of the screen. Select Column chart in the Charts group. Select the leftmost figure in the 2-D Column row. If you would like, add horizontal and vertical axis titles.

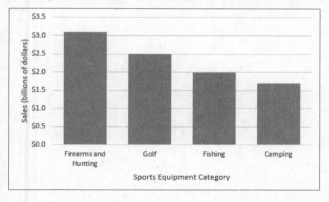

Cumulative Review Exercises

1.   Pareto chart.

2.   Nominal, because the responses consist of names only. The responses do not measure or count anything, and they cannot be arranged in order according to some quantitative scale.

3.   Voluntary response sample. The voluntary response sample is not likely to be representative of the population, because those with special interests or strong feelings about the topic are more likely than others to respond and their views might be very different from those of the general population.

4.   By using a vertical scale that does not begin at 0, the graph exaggerates the differences in the numbers of responses. The graph could be modified by starting the vertical scale at 0 instead of 50.

5.   The percentage is $\frac{241}{641} = 0.376 = 37.6\%$. Because the percentage is based on a sample and not a population that percentage is a statistic.

6.   Enter the label **Minutes** in cell A1 of an Excel worksheet followed by the times data given in Exercise 6. Select the **XLSTAT** add-in. Select **Describing data**. Select **Histograms**. Enter the range of the times data including the label (A1:A21). Select **Discrete**. Select **Sample labels**. Click the **Options** tab. Select **Range** and enter **10**. Click the **Charts** tab. Select **Histograms** and **Bars**. Select **Frequency** for the ordinate of the histogram. Click **OK**. The first three columns of the frequency distribution output are shown below.

Lower bound	Upper bound	Frequency
0	10	2
10	20	3
20	30	9
30	40	4
40	50	2
50	60	0

7.   The histogram was included in the output for Exercise 6. Because the frequencies increase to a maximum and then decrease and the left half of the histogram is roughly a mirror image of the right half, the data appear to be from a population with a normal distribution.

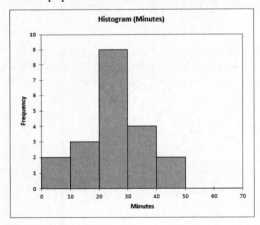

8.   Stemplot

     0 | 0 5

     1 | 2 5 5

     2 | 0 2 4 5 5 5 7 7 8

     3 | 0 0 5 5

     4 | 0 5

Chapter 3

Statistics for Describing, Exploring, and Comparing Data

Section 3-2, Basic Skills and Concepts

1. No. The numbers do not measure or count anything, so the mean would be a meaningless statistic.

3. No. The price exactly in between the highest and lowest is the midrange, not the median.

5. Enter the top 10 celebrity incomes in an Excel worksheet. Click on **DATA**. Select **Data Analysis** and select **Descriptive Statistics**. Enter the data range and select **Summary statistics**. Click **OK**.

   The mean is $159.8 million.

   The median is $95 million.

   There is no mode.

   The midrange is $\dfrac{332+67}{2}=\$199.5$ million.

   Apart from the obvious and trivial fact that the mean annual earnings of all celebrities is less than $332 million, nothing meaningful can be known about the mean of the population.

Income	
Mean	159.8
Standard Error	32.47762
Median	95
Mode	#N/A
Standard Deviation	102.7032
Sample Variance	10547.96
Kurtosis	-1.27867
Skewness	0.758413
Range	265
Minimum	67
Maximum	332
Sum	1598
Count	10

7. Enter the head injury measurements in an Excel worksheet. Click on **DATA**. Select **Data Analysis** and select **Descriptive Statistics**. Enter the data range and select **Summary statistics**. Click **OK**.

   The mean is 430.1 hic.

   The median is 393 hic.

   There is no mode.

   The midrange is $\dfrac{544+326}{2}=435$ hic.

   The safest of these cars appears to be the Hyndai Elantra. Because the measurements appear to vary substantially from a low of 326 hic to a high of 544 hic, it appears that some small cars are considerably safer than others.

hic	
Mean	430.1429
Standard Error	33.55126
Median	393
Mode	#N/A
Standard Deviation	88.76829
Sample Variance	7879.81
Kurtosis	-2.24005
Skewness	0.254214
Range	218
Minimum	326
Maximum	544
Sum	3011
Count	7

9. Enter the gross amounts earned in an Excel worksheet. Click on **DATA**. Select **Data Analysis** and select **Descriptive Statistics**. Enter the data range and select **Summary statistics**. Click **OK**.

   The mean is $16.4 million.

   The median is $10 million.

   Excel's Descriptive Statistics output only reports one mode even when the distribution is multimodal. To check for multiple modes, sort the data in either ascending or descending order. This distribution has three modes: $4 million, $9 million, and $10 million.

   The midrange is $\dfrac{58+4}{2}=\$31$ million.

Millions of dollars	
Mean	16.35714
Standard Error	3.88093
Median	10
Mode	10
Standard Deviation	14.52111
Sample Variance	210.8626
Kurtosis	4.735902
Skewness	2.02466
Range	54
Minimum	4
Maximum	58
Sum	229
Count	14

The measures of center do not reveal anything about the pattern of data over time, and that pattern is a key component of a movie's success. The first amount is highest for the opening day when many Harry Potter fans are most eager to see the movie, the third and fourth values are from the first Friday and the first Saturday, which are the popular weekend days when movie attendance tends to spike.

11. Enter the software prices in an Excel worksheet. Click on **DATA**. Select **Data Analysis** and select **Descriptive Statistics**. Enter the data range and select **Summary statistics**. Click **OK**.

The mean is $59.22.

The median is $57.84

There is no mode.

The midrange is $\dfrac{71.77+48.92}{2}=\$60.345$.

None of the measures of center are most important here. The most relevant statistic in this case is the minimum value of $48.92, because that is the lowest price for the software. Here, we generally care about the lowest price not the

mean price or the median price.

Software price	
Mean	59.21667
Standard Error	4.064902
Median	57.835
Mode	#N/A
Standard Deviation	9.956935
Sample Variance	99.14055
Kurtosis	-1.92249
Skewness	0.31039
Range	22.85
Minimum	48.92
Maximum	71.77
Sum	355.3
Count	6

13. Enter the lead concentrations in an Excel worksheet. Click on **DATA**. Select **Data Analysis** and select **Descriptive Statistics**. Enter the data range and select **Summary statistics**. Click **OK**.

The mean is 11.05 ug/g.

The median is 9.5 ug/g.

Excel's Descriptive Statistics output only reports one mode even when the distribution is multimodal. To check for multiple modes, sort the data in either ascending or descending order. The distribution is unimodal. The mode is 20.5 ug/g.

The midrange is $\dfrac{20.5+3}{2}=11.75$ ug/g.

ug/g	
Mean	11.05
Standard Error	2.043214
Median	9.5
Mode	20.5
Standard Deviation	6.461209
Sample Variance	41.74722
Kurtosis	-1.32533
Skewness	0.513238
Range	17.5
Minimum	3
Maximum	20.5
Sum	110.5
Count	10

There is not enough information given here to assess the true danger of these drugs, but ingestion of any lead is generally detrimental to good health. All of the decimal values are either 0 or 5, so it appears that the lead concentrations were rounded to the nearest one-half unit of measurement(ug/g).

15. Enter the number of years data in an Excel worksheet. Click on **DATA**. Select **Data Analysis** and select **Descriptive Statistics**. Enter the data range and select **Summary statistics**. Click **OK**.

The mean is 6.5 years.

The median is 4.5 years.

The distribution is bimodal. The two modes are 4 years and 4.5 years.

The midrange is $\dfrac{15+4}{2}=9.5$ years.

It is common to earn a bachelor's degree in four years, but the typical college student requires more than four years.

Years	
Mean	6.5
Standard Error	0.783884
Median	4.5
Mode	4
Standard Deviation	3.505635
Sample Variance	12.28947
Kurtosis	0.863072
Skewness	1.433402
Range	11
Minimum	4
Maximum	15
Sum	130
Count	20

17. Enter the arrival delay times in an Excel worksheet. Click on **DATA**. Select **Data Analysis** and select **Descriptive Statistics**. Enter the data range and select **Summary statistics**. Click **OK**.

The mean is –14.3 min.

The median is –16.5 min.

The mode is –32 min.

The midrange is $\frac{11+(-32)}{2} = -10.5$ min.

Because the measures of center are all negative values, it appears that the flights tend to arrive early before the scheduled arrival times, so the on-time performance appears to be very good.

Minutes	
Mean	-14.25
Standard Error	5.377699
Median	-16.5
Mode	-32
Standard Deviation	15.21043
Sample Variance	231.3571
Kurtosis	-0.55855
Skewness	0.486537
Range	43
Minimum	-32
Maximum	11
Sum	-114
Count	8

19. Enter the jersey numbers in an Excel worksheet. Click on **DATA**. Select **Data Analysis** and select **Descriptive Statistics**. Enter the data range and select **Summary statistics**. Click **OK**.

The mean is 50.4

The median is 73.

There is no mode.

The midrange is $\frac{78+9}{2} = 48.5$.

The numbers do not measure or count anything; they are simply replacements for names. The data are at the nominal level of measurement, and it makes no sense to compute the measures of center for these data.

Jersey No.	
Mean	50.36364
Standard Error	9.618422
Median	73
Mode	#N/A
Standard Deviation	31.9007
Sample Variance	1017.655
Kurtosis	-2.1954
Skewness	-0.22559
Range	79
Minimum	9
Maximum	88
Sum	554
Count	11

21. Enter the miles per hour data of the white drivers and the African American drivers in adjacent columns of an Excel worksheet. Click on **DATA**. Select **Data Analysis** and select **Descriptive Statistics**. Enter the data range inclusive of both white and African American drivers and select **Summary statistics**. Click **OK**.

White drivers' mean is 73 mph.

White drivers' median is 73 mph.

African American drivers' mean is 74 mph.

African American drivers' median is 74 mph.

Although the African American drivers have a mean speed greater than the white drivers, the difference is very small, so it appears that drivers of both races speed about the same amount.

	White	African American
Mean	73	74
Standard Error	0.918937	0.869227
Median	73	74
Mode	74	74
Standard Deviation	2.905933	2.7487371
Sample Variance	8.444444	7.5555556
Kurtosis	-1.09495	-0.2255
Skewness	3.08E-17	0.1605015
Range	8	9
Minimum	69	70
Maximum	77	79
Sum	730	740
Count	10	10

23. Enter the contributions (in dollars) made to the Obama and McCain campaigns in adjacent columns of an Excel worksheet. Click on **DATA**. Select **Data Analysis** and select **Descriptive Statistics**. Enter the data range inclusive of both the Obama and McCain data and **Select Summary statistics**. Click **OK**.

Obama had a mean of $653.9 and a median of $452.

McCain had a mean of $458.5 and a median of $350.

The contributions appear to favor Obama because his mean and median are substantially higher.

With 66 contributions to Obama and 20 to McCain, Obama collected substantially more in total number of contributions.

	Obama	McCain
Mean	653.8667	458.5333
Standard Error	135.0921	108.0361
Median	452	350
Mode	1000	500
Standard Deviation	523.2096	418.422
Sample Variance	273748.3	175077
Kurtosis	1.610286	2.184052
Skewness	1.282105	1.579109
Range	1900	1460
Minimum	100	40
Maximum	2000	1500
Sum	9808	6878
Count	15	15

25. Open the **QUAKE** data file on your data disk. Click on **DATA**. Select **Data Analysis** and select **Descriptive Statistics**. Enter the data range of the earthquake magnitudes including the label MAG (**A1:A51**). Select **Labels in First Row** and select **Summary statistics**. Click **OK**.

The mean is 1.184 and the median is 1.235. Yes, it is an outlier because it is a value that is very far away from all the other sample values.

	MAG
Mean	1.1842
Standard Error	0.083059
Median	1.235
Mode	1.25
Standard Deviation	0.587315
Sample Variance	0.344939
Kurtosis	0.822487
Skewness	0.64086
Range	2.95
Minimum	0
Maximum	2.95
Sum	59.21
Count	50

27. Open the **POTUS** data file on your data disk. Click on **DATA**. Select **Data Analysis** and select **Descriptive Statistics**. Enter the data range of the number of years that U.S. presidents have lived after their first inauguration, including the label Years (**D1:D35**). Select **Labels in First Row** and select **Summary statistics**. Click **OK**.

The mean is 15.0 years and the median is 16 years. Presidents receive Secret Service protection after they leave office, so the mean is helpful in planning for cost and resources used for that protection.

	Years
Mean	15.0303
Standard Error	1.69194
Median	16
Mode	4
Standard Deviation	9.719455
Sample Variance	94.4678
Kurtosis	-0.86558
Skewness	0.121597
Range	36
Minimum	0
Maximum	36
Sum	496
Count	33

29. The x values are the class midpoints from the frequency distribution given in the problem, the f values are the frequencies, and f*x values are the products.

x	f	f*x
24.5	27	661.5
34.5	34	1173
44.5	13	578.5
54.5	2	109
64.5	4	258
74.5	1	74.5
84.5	1	84.5
	82	2939

The mean = $\sum (f \cdot x). \sum f = 2939 / 82 = 35.8$. This result is quite close to the mean of 35.9 years found by using the original list of data values.

31. The x values are the class midpoints from the frequency distribution given in the problem, the f values are the frequencies, and the f*x values are the products.

x	f	f*x
54.5	4	218
64.5	10	645
74.5	25	1862.5
84.5	43	3633.5
94.5	26	2457
104.5	8	836
114.5	3	343.5
124.5	2	249
	121	10244.5

The mean = $\sum(f \cdot x).\sum f = 10244.5/12 = 84.7$. This result is close to the mean of 84.4 found by using the original list of data values.

Section 3-2, Beyond the Basics

33. a. $x = 5(0.62) - 0.3 - 0.4 - 1.1 - 0.7 = 0.6$ parts per million

  b. $n-1$

35. The mean is 39.070, the 10% trimmed mean is 27.677, and the 20% trimmed mean is 27.176. By deleting the outlier of 472.2, the trimmed means are substantially different from the untrimmed mean.

37. The geometric mean si $\sqrt[5]{1.017 \times 1.037 \times 1.052 \times 1.051 \times 1.027} = 1.036711036$, or 1.0367 when rounded. Single percentage growth rate is 3.67%. The result is not exactly the same as the mean which is 3.68%.

39. The median is $30 + (10)\dfrac{\dfrac{27+34+13+2+4+1+1+1}{2} - (27+1)}{34} = 33.970588$ years, which is rounded to 34 years.

The value of 33 years is better because it is based on the original data and does not involve interpolation.

Section 3-3, Basic Skills and Concepts

1. The IQ scores of a class of statistics students should have less variation, because those students are a much more homogeneous group with IQ scores that are likely to be closer together.

3. Variation is a general descriptive term that refers to the amount of dispersion or spread among the data values, but the variance refers specifically to the square of the standard deviation.

5. For some of the problems in Section 3-2, you used Excel's Data Analysis Tools to find summary measures of a distribution. For some of the problems in Section 3-3, you will use XLSTAT. Type the label **Millions of dollars** in an Excel worksheet followed by the earnings of 10 celebrities. Select the **XLSTAT** add-in. Select **Describing data**. Select **Descriptive statistics**. Below Quantitative data, enter the data range inclusive of the label. Select **Sample labels**. Click the **Options** tab. Select **Descriptive Statistics**. Click the **Outputs** tab. Select **Range**, **Variance (n-1)**, and **Standard deviation (n-1)**. Click **OK**.

The range is $265 million.

The variance is 10,547.956 square of million dollars.

The standard deviation is $102.703 million.

Statistic	Millions of dollars
Range	265.0000
Variance (n-1)	10547.9556
Standard deviation (n-1)	102.7032

Because the data values are the 10 highest from the population, nothing meaningful can be known about the standard deviation of the population.

7. Type the label **hic** in an Excel worksheet followed by the seven measurements given in the problem. Select the **XLSTAT** add-in. Select **Describing data**. Select **Descriptive statistics**. Below Quantitative data, enter the data range inclusive of the label. Select **Sample labels**. Click the **Options** tab. Select **Descriptive Statistics**. Click the **Outputs** tab. Select **Range**, **Variance (n-1)**, and **Standard deviation (n-1)**. Click **OK**.

The range is 218 hic.

The variance is 7,879.8 hic squared.

The standard deviation is 88.8 hic.

Statistic	hic
Range	218.0000
Variance (n-1)	7879.8095
Standard deviation (n-1)	88.7683

Although all of the cars are small, the range from 326 to 544 hic appears to be relatively large, so the head injury measurements are not about the same.

9. Type the label **Millions of dollars** in an Excel worksheet followed by the gross amounts earned given in the problem. Select the **XLSTAT** add-in. Select **Describing data**. Select **Descriptive statistics**. Below Quantitative data, enter the data range inclusive of the label. Select **Sample labels**. Click the **Options** tab. Select **Descriptive Statistics**. Click the **Outputs** tab. Select **Range**, **Variance (n-1)**, and **Standard deviation (n-1)**. Click **OK**.

The range is $54 million.

The variance is 210.9 square of million dollars.

The standard deviation is $14.5 million.

Statistic	Millions of dollars
Range	54.0000
Variance (n-1)	210.8626
Standard deviation (n-1)	14.5211

An investor would care about the gross from opening day and the rate of decline after that, but the measures of center and variation are less important.

11. Type the label **Price** in an Excel worksheet followed by the data given in the problem. Select the **XLSTAT** add-in. Select **Describing data**. Select **Descriptive statistics**. Below Quantitative data, enter the data range inclusive of the label. Select **Sample labels**. Click the **Options** tab. Select **Descriptive Statistics**. Click the **Outputs** tab. Select **Range**, **Variance (n-1)**, and **Standard deviation (n-1)**. Click **OK**.

The range is $22.85.

The variance is 99.141 dollars squared.

The standard deviation is $9.957.

Statistic	Price
Range	22.8500
Variance (n-1)	99.1405
Standard deviation (n-1)	9.9569

The measures of variation are not very helpful in trying to find the best deal.

13. Type the label **ug/g** in an Excel worksheet followed by the data given in the problem. Select the **XLSTAT** add-in. Select **Describing data**. Select **Descriptive statistics**. Below Quantitative data, enter the data range inclusive of the label. Select **Sample labels**. Click the **Options** tab. Select **Descriptive Statistics**. Click the **Outputs** tab. Select **Range**, **Variance (n-1)**, and **Standard deviation (n-1)**. Click **OK**.

The range is 17.50 ug/g.

The variance is 41.75 ug/g squared.

The standard deviation is 6.46 ug/g.

Statistic	ug/g
Range	17.5000
Variance (n-1)	41.7472
Standard deviation (n-1)	6.4612

If the medicines contained no lead all of the measures would be 0 ug/g, and the measures of variation would all be 0 as well.

15. Type the label **Years** in an Excel worksheet followed by the data given in the problem. Select the **XLSTAT** add-in. Select **Describing data**. Select **Descriptive statistics**. Below Quantitative data, enter the data range inclusive of the label. Select **Sample labels**. Click the **Options** tab. Select **Descriptive Statistics**. Click the **Outputs** tab. Select **Range**, **Variance (n-1)**, and **Standard deviation (n-1)**. Click **OK**.

The range is 11 years.

The variance is 12.3 years squared.

The standard deviation is 3.5 years.

Statistic	Years
Range	11.0000
Variance (n-1)	12.2895
Standard deviation (n-1)	3.5056

No, because 12 years is within 2 standard deviations of the mean.

17. Type the label **Minutes** in an Excel worksheet followed by the data given in the problem. Select the **XLSTAT** add-in. Select **Describing data**. Select **Descriptive statistics**. Below Quantitative data, enter the data range inclusive of the label. Select **Sample labels**. Click the **Options** tab. Select **Descriptive Statistics**. Click the **Outputs** tab. Select **Range**, **Variance (n-1)**, and **Standard deviation (n-1)**. Click **OK**.

The range is 43 minutes.

The variance is 231.4 minutes squared.

The standard deviation is 15.2 minutes.

Statistic	Minutes
Range	43.0000
Variance (n-1)	231.3571
Standard deviation (n-1)	15.2104

No. The standard deviation can never be negative.

19. Type the label **Jersey No.** in an Excel worksheet followed by the data given in the problem. Select the **XLSTAT** add-in. Select **Describing data**. Select **Descriptive statistics**. Below Quantitative data, enter the data range inclusive of the label. Select **Sample labels**. Click the **Options** tab. Select **Descriptive Statistics**. Click the **Outputs** tab. Select **Range**, **Variance (n-1)**, and **Standard deviation (n-1)**. Click **OK**.

The range is 79.

The variance is 1017.7.

The standard deviation is 31.9.

Statistic	Jersey No.
Range	79.0000
Variance (n-1)	1017.6545
Standard deviation (n-1)	31.9007

The data are at the nominal level of measurement so it makes no sense to compute the measures of variation for these data.

21. Type the labels **White** and **African American** in the top cells of adjacent columns in an Excel worksheet. Then enter each group's mi/h data under their labels. Select the **XLSTAT** add-in. Select **Describing data**. Select **Descriptive statistics**. Below Quantitative data, enter the data range inclusive of the white data, the African American data, and the labels. Select **Sample labels**. Click the **Options** tab. Select **Descriptive Statistics**. Click the **Outputs** tab. Select **Mean** and **Standard deviation (n-1)**. Click **OK**.

The mean of the White drivers is 73.00 and the standard deviation is 2.906. The coefficient of variation for the White drivers is $\frac{2.906}{73} \cdot 100 = 4\%$. The mean of the African American drivers is 74.00 and the standard deviation is 2.749. The coefficient of variation for the African American drivers is $\frac{2.749}{74} \cdot 100 = 3.7\%$. The variation is about the same.

Statistic	White	African American
Mean	73.0000	74.0000
Standard deviation (n-1)	2.9059	2.7487

23. Type the labels **Obama** and **McCain** in the top cells of adjacent columns in an Excel worksheet. Then enter each candidate's data under his label. Select the **XLSTAT** add-in. Select **Describing data**. Select **Descriptive statistics**. Below Quantitative data, enter the data range inclusive of both the Obama and McCain data and the labels. Select **Sample labels**. Click the **Options** tab. Select **Descriptive Statistics**. Click the **Outputs** tab. Select **Mean** and **Standard deviation (n-1)**. Click **OK**.

The mean of Obama's contribution is $654 and the standard deviation is $523.2. The coefficient of variation for Obama is $\frac{523.2}{654} \cdot 100 = 80\%$. The mean for McCain is $459 and the

Statistic	Obama	McCain
Mean	653.8667	458.5333
Standard deviation (n-1)	523.2096	418.4220

standard deviation is $418.4. The coefficient of variation for McCain is $\frac{418.4}{459} \cdot 100 = 91\%$. The variation among Obama contributors is a little less than the variation among the McCain contributors.

25. Open the **QUAKE** data file on your data disk. Select the **XLSTAT** add-in. Select **Describing data**. Select **Descriptive statistics**. Below Quantitative data, enter the data range of the earthquake magnitudes including the label MAG (**A1:A51**). Select **Sample labels**. Click the **Options** tab. Select

Statistic	MAG
Range	2.9500
Variance (n-1)	0.3449
Standard deviation (n-1)	0.5873

Descriptive Statistics. Click the **Outputs** tab. Select **Range**, **Variance (n-1)**, and **Standard deviation (n-1)**. Click **OK**. The range is 2.95, the variance is 0.345, and the standard deviation is 0.587.

27. Open the **POTUS** data file on your data disk. Select the **XLSTAT** add-in. Select **Describing data**. Select **Descriptive statistics**. Below Quantitative data, enter the data range of the number of years that U.S. presidents have lived after their first inauguration, including the label Years (**D1:D35**). Select **Sample labels**. Click the **Options** tab. Select **Descriptive Statistics**. Click the **Outputs** tab. Select **Range**, **Variance (n-1)**, and **Standard deviation (n-1)**. Click **OK**. The range is 36 years, the variance is 94.5 years squared, and the standard deviation is 9.7 years.

Statistic	Years
Range	36.0000
Variance (n-1)	94.4678
Standard deviation (n-1)	9.7195

29. The standard deviation $\dfrac{2.95}{4} = 0.738$, which is not substantially different from 0.587.

31. The standard deviation $\dfrac{36}{4} = 9$ years, which is reasonably close to 9.7 years.

33. No. The pulse rate of 99 beats per minute is between the minimum usual value of 54.3 beats per minute and the maximum usual value of 100.7 beats per minute.

35. Yes. The volume of 11.9 oz. is not between the minimum usual value of 11.97 oz. and the maximum usual value of 12.41 oz.

37. $s = \sqrt{\dfrac{82(84,408.5) - 8,637,721}{82(81)}} = 12.3$ years. This result is not substantially different from the standard deviation of 11.1 years found from the original list of data values.

39. $s = \sqrt{\dfrac{121(889,106.69) - 104,941,584.81}{121(120)}} = 13.5$. The result is very close to the standard deviation of 13.4 found from the original list of sample values.

41. a.   95%

    b.   68%

43. At least 75% of women have platelet counts within 2 standard deviations of the mean. The minimum is 150 and the maximum is 410.

Section 3-3, Beyond the Basics

45. a. $\sigma^2 = \dfrac{(2-4.33)^2 + (3-4.33)^2 + (8-4.33)^2}{3} = 6.9 \, \text{min}^2$

    b.   The nine possible samples of two values are the following: [(2 min, 2 min), (2 min, 3 min), (2 min, 8 min), (3 min, 2 min), (3 min, 3 min), (3 min, 8 min), (8 min, 2 min), (8 min, 3 min), (8 min, 8 min)] and they have the following corresponding variances: [0, 0.707, 18, 0.707, 0, 12.5, 18, 12.5, 0] which have the mean of 6.934.

    c.   The population variances of the nine samples above are [0, 0.3535, 9, 0.3535, 0, 6.25, 9, 6.25, 0]. The mean is 3.4 min. squared.

    d.   Part (b), because repeated samples result in variances that target the same value (6.9 min.²) as the population variance. Use division by $n-1$.

    e.   No. The mean of the sample variances (6.9 min.²) equals the population variance, but the mean of the sample standard deviations (1.9 min.) does not equal the mean of the population standard deviation (2.6 min.)

Section 3-4, Basic Skills and Concepts

1. Madison's height is below the mean. It is 2.28 standard deviations below the mean.

3. The lowest amount is $5 million, the first quartile Q_1 is $47 million, the second quartile Q_2 (or median) is $104 million, the third quartile Q_3 is $121 million, and the highest gross amount is $380 million.

5. a.   The difference is $\$3,670,505 - \$4,939,455 = -\$1,268,950$

    b. $\dfrac{\$1,268,950}{\$7,775,948} = 0.16$ standard deviations

    c.   You will use Excel's STANDARDIZE function to solve the problem. Click **FORMULAS** at the top of the screen, select **Insert Function**, select the **Statistical** category, select the **STANDARDIZE** function, and click **OK**. Enter the following values: x = 3670505, Mean = 4939455, and Standard dev = 7775948. Click OK. The function returns $z = -0.16$.

    d.   Usual

7. a.   The difference is $\$1 - \$1,449,779 = -\$1,449,778$

    b. $\dfrac{\$1,449,778}{\$527,651} = 2.75$ standard deviation

    c.   You will use Excel's STANDARDIZE function to solve the problem. Click **FORMULAS** at the top of the screen, select **Insert Function**, select the **Statistical** category, select the **STANDARDIZE** function, and click **OK**. Enter the following values: x = 1, Mean = 1449779, and Standard dev = 527651. Click OK. The function returns $z = -2.75$.

    d.   Unusual

9. z scores of –2 and 2. A z score of –2 means a score of $x = -2 \cdot 15 + 100 = 70$. A z score of 2 means a score of $x = 2 \cdot 15 + 100 = 130$

11. Two standard deviations from the mean: $1.240 - 2 \cdot 0.578 = 0.084$ and $1.240 + 2 \cdot 0.578 = 2.396$

13. You will use Excel's STANDARDIZE function to solve the problem. Click **FORMULAS** at the top of the screen, select **Insert Function**, select the **Statistical** category, select the **STANDARDIZE** function, and click **OK**. For Sultan Kosen, enter the following values: x = 247, Mean = 175, and Standard dev = 7. Click OK. The function returns $z = 10.29$. Repeat the procedure, except this time, for De-Fen Yao, enter the following values: x = 236, Mean = 162, and Standard dev = 6. Click **OK**. The function returns $z = 12.33$. De-Fen Yao is relatively taller, because her z score of 12.33 is greater than the z score of 10.29 for Sultan Kosen. De-Fen Yao is more standard deviations above the mean than Sultan Kosen.

15. You will use Excel's STANDARDIZE function to solve the problem. Click **FORMULAS** at the top of the screen, select **Insert Function**, select the **Statistical** category, select the **STANDARDIZE** function, and click **OK**. For the SAT, enter the following values: x = 1490, Mean = 1518, and Standard dev = 325. Click OK. The function returns $z = -0.09$. Repeat the procedure, except this time, for the ACT, enter the following values: x = 17, Mean = 21.1, and Standard dev = 4.8. Click **OK**. The function returns $z = -0.85$. The z score of –0.09 is higher than the z score of –0.85, so the SAT score of 1490 is relatively better.

17. You will use Excel's PERCENTRANK.INC function to solve the problem. Enter the data in column A of an Excel worksheet, A1:A24. Click **FORMULAS** at the top of the screen, select **Insert Function**, select the **Statistical** category, select the **PERCENTRANK.INC** function, and click **OK**. Enter the following information: Array = A1:A24, x = 213, and Significance = 2. Click **OK**. The function returns 0.13. The percentile rank of 213 is $0.13 \cdot 100$, the 13th percentile.

19. You will use Excel's PERCENTRANK.INC function to solve the problem. Enter the data in column A of an Excel worksheet, A1:A24. Click **FORMULAS** at the top of the screen, select **Insert Function**, select the **Statistical** category, select the **PERCENTRANK.INC** function, and click **OK**. Enter the following information: Array = A1:A24, x = 250, and Significance = 2. Click **OK**. The function returns 0.52. The percentile rank of 250 is $0.52 \cdot 100$, the 52nd percentile.

21. You will use Excel's PERCENTILE.INC function to solve the problem. Enter the data in column A of an Excel worksheet, A1:A24. Click **FORMULAS** at the top of the screen, select **Insert Function**, select the **Statistical** category, select the **PERCENTILE.INC** function, and click **OK**. Enter the following information: Array = A1:A24, k = 0.60. Click **OK**. The function returns 250.8. P_{60} = 250.8.

23. You will use Excel's QUARTILE.INC function to solve the problem. Enter the data in column A of an Excel worksheet, A1:A24. Click **FORMULAS** at the top of the screen, select **Insert Function**, select the **Statistical** category, select the **QUARTILE.INC** function, and click **OK**. Enter the following information: Array = A1:A24, Quart = 3. Click **OK**. The function returns 255. Q_3 = 255.

25. You will use Excel's PERCENTILE.INC function to solve the problem. Enter the data in column A of an Excel worksheet, A1:A24. Click **FORMULAS** at the top of the screen, select **Insert Function**, select the **Statistical** category, select the **PERCENTILE.INC** function, and click **OK**. Enter the following information: Array = A1:A24, k = 0.50. Click **OK**. The function returns 247.5. P_{50} = 247.5.

27. You will use Excel's PERCENTILE.INC function to solve the problem. Enter the data in column A of an Excel worksheet, A1:A24. Click **FORMULAS** at the top of the screen, select **Insert Function**, select the **Statistical** category, select the **PERCENTILE.INC** function, and click **OK**. Enter the following information: Array = A1:A24, k = 0.25. Click **OK**. The function returns 234.75. P_{25} = 234.75.

29. You will use XLSTAT to solve the problem. Type the label **Minutes** in an Excel worksheet followed by the duration times of all missions flown by the space shuttle *Challenger*. Select the **XLSTAT** add-in. Select **Describing data**. Select **Descriptive statistics**. Below Quantitative data, enter the data range inclusive of the label. Select **Sample labels**. Click the **Options** tab. Select **Descriptive Statistics** and **Charts**. Click the **Outputs** tab. Select **Minimum**, **Maximum**, **1st Quartile**, **Median**, and **3rd Quartile**. Click the **Charts (1)** tab. Select **Box plots**. Click **OK**. The modified boxplot is displayed at the right and the 5-number summary is displayed below.

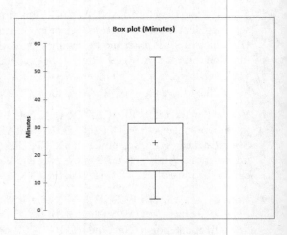

Statistic	Minutes
Minimum	1.0000
Maximum	11844.0000
1st Quartile	8727.7500
Median	10074.5000
3rd Quartile	11115.0000

31. You will use XLSTAT to solve the problem. Type the label **Minutes** in an Excel worksheet followed by the interval times. Select the **XLSTAT** add-in. Select **Describing data**. Select **Descriptive statistics**. Below Quantitative data, enter the data range inclusive of the label. Select **Sample labels**. Click the **Options** tab. Select **Descriptive Statistics** and **Charts**. Click the **Outputs** tab. Select **Minimum**, **Maximum**, **1st Quartile**, **Median**, and **3rd Quartile**. Click the **Charts (1)** tab. Select **Box plots**. Click **OK**. The modified boxplot is displayed at the right and the 5-number summary is displayed below.

Statistic	Minutes
Minimum	4.0000
Maximum	63.0000
1st Quartile	14.2500
Median	18.0000
3rd Quartile	31.5000

33. Open the **MBODY** data file on your data disk. Select the **XLSTAT** add-in. Select **Describing data**. Select **Descriptive statistics**. Below Quantitative data, enter the range of the males' PULSE data inclusive of the label. Select **Sample labels**. Click the **Options** tab. Select **Charts**. Click the **Charts (1)** tab. Select **Box plots**. Select **Horizontal**. Click **OK**. The modified boxplot for the males' data is displayed below on the left.

Open the **FBODY** data file on your data disk. Select the **XLSTAT** add-in. Select **Describing data**. Select **Descriptive statistics**. Below Quantitative data, enter the range of the females' PULSE data inclusive of the label. Select **Sample labels**. Click the **Options** tab. Select **Charts**. Click the **Charts (1)** tab. Select **Box plots**. Select **Horizontal**. Click **OK**. The modified boxplot for the females' data is displayed at the right.

It appears that males have lower pulse rates than females.

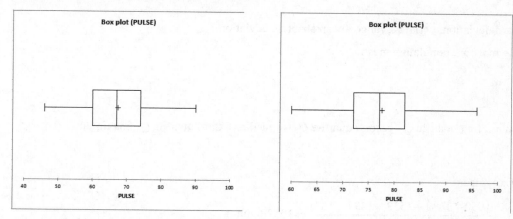

35. Open the **COLA** data file on your data disk. Copy and paste the CKREGWT and CKDIETWT data so that they are in adjacent columns in an Excel worksheet. Select the **XLSTAT** add-in. Select **Describing data**. Select **Descriptive statistics**. Below Quantitative data, enter the range of both CKWEGWT and CKDIETWT data, inclusive of the labels. Select **Sample labels**. Click the **Options** tab. Select **Charts**. Click the **Charts (1)** tab. Select **Box plots**. Select **Horizontal**. Click **OK**. The modified boxplots are displayed below. The weights of regular Coke appear to be generally greater than those of diet Coke, probably due to the sugar in the cans of regular coke.

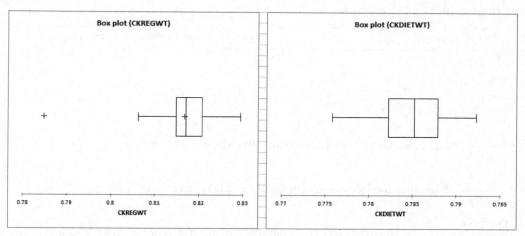

Section 3-4, Beyond the Basics

37. XLSTAT generates only modified boxplots. Follow the instructions given in Exercise 34.

Chapter Quick Quiz

1. The mean is 14.0 minutes

2. The median is 12 minutes

3. The mode is 12 minutes

4. The variance is $(5 \text{ min})^2 = 25 \text{ min}^2$

5. $z = \dfrac{6-11.4}{7} = -0.77$

6. Standard deviation, variance, range, mean absolute deviation

7. Sample mean $\overline{x}$, population mean μ

8. s, σ, s^2, σ^2

9. 75%

10. Minimum, first quartile Q_1, second quartile Q_2 (or median), third quartile Q_3, maximum

Review Exercises

1. a. $\overline{x} = \dfrac{1550+1642+1538+1497+1571}{5} = 1559.6 \text{ mm}$

   b. The median is 1550 mm

   c. There is no mode

   d. The midrange is $\dfrac{1497+1642}{2} = 1569.5 \text{ mm}$

   e. The range is $1642 - 1497 = 145 \text{ mm}$

   f. $s = \sqrt{\dfrac{(1550-1559.6)^2 + (1642-1559.6)^2 + (1538-1559.6)^2 + (1497-1559.6)^2 + (1571-1559.6)^2}{5-1}}$

   $= 53.37$ mm

   g. $s^2 = 53.37^2 = 2849.3 \text{ mm}^2$

   h. $Q_1 = \dfrac{25 \cdot 5}{100} = 1.25$, pick second entry (in ordered list) which is 1538 mm

   i. $Q_3 = \dfrac{75 \cdot 5}{100} = 3.75$, pick the fourth entry (in the ordered list) which is 1571 mm

2. $z = \dfrac{1642-1559.6}{53.37} = 1.54$. The eye height is not unusual because its z score is between –2 and 2, so it is within two standard deviations of the mean.

3. You will use XLSTAT to solve the problem. Type the label **Millimeters** in an Excel worksheet followed by the five data values. Select the **XLSTAT** add-in. Select **Describing data**. Select **Descriptive statistics**. Below Quantitative data, enter the data range inclusive of the label. Select **Sample labels**. Click the **Options** tab. Select **Descriptive Statistics** and **Charts**. Click the **Outputs** tab. Select **Minimum, Maximum, 1st Quartile, Median**, and **3rd Quartile**. Click the **Charts (1)** tab. Select **Box plots**. Select **Horizontal**. Click **OK**. The modified boxplot and the 5-number summary are displayed at the top of the next page.

Statistic	Millimeters
Minimum	1497.0000
Maximum	1642.0000
1st Quartile	1538.0000
Median	1550.0000
3rd Quartile	1571.0000

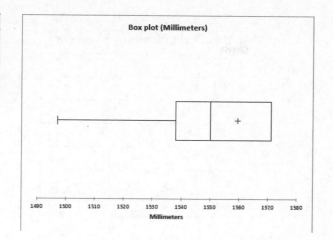

Because the boxplot shows a distribution of data that is roughly symmetric, the data could be from a population with a normal distribution, but the data are not necessarily from a population with a normal distribution, because there is no way to determine whether a histogram is roughly bell-shaped.

4. The mean is 10053.5. The ZIP codes do not measure or count anything. They are at the nominal level of measurement, so the mean is a meaningless statistic.

5. The male z score is $z = \dfrac{28 - 26.601}{5.359} = 0.26$. The female z score is $z = \dfrac{29 - 28.441}{7.394} = 0.08$. The male has a larger relative BMI because the male has the larger z score.

6. a.   The answers may vary but a mean around $8 or $9 is reasonable.

    b.   A reasonable standard deviation would be around $1 or $2.

7. Based on a minimum age of 23 years and a maximum age of 70 years an estimate of the age standard deviation would be $\dfrac{70 - 23}{4} = 11.75$ years.

8. A minimum usual sitting height of $914 - 2 \cdot 36 = 842$ mm and a maximum sitting height of $914 + 2 \cdot 36 = 986$ mm. The maximum usual height of 986 mm is more relevant for designing overhead bin storage.

9. The minimum value is 963 cm^3, the first quartile is 1034.5 cm^3, the second quartile (or median) is 1079 cm^3, the third quartile is 1188.5 cm^3, and the maximum value is 1439 cm^3.

10. The median would be better because it is not affected much by the one very large income.

Cumulative Review Exercises

1. a.   Continuous

    b.   Ratio

2.

Hand Length (mm)	Frequency
150 – 159	1
160 – 169	0
170 – 179	2
180 – 189	0
190 – 199	3
200 – 209	1
210 – 219	1

3.   Hand length histogram

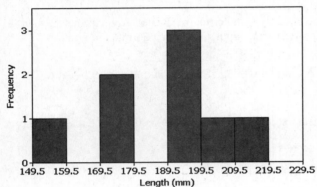

4.

15 | 8

16 |

17 | 3 9

18 |

19 | 5 6 9

20 | 7

21 | 4

5.   a. $\bar{x} = \dfrac{173+179+207+158+196+195+214+199}{8} = 190.1\,\text{mm}$

   b.   The median is 195.5 mm

   c. $(173-190.1)^2 + (179-190.1)^2 + (207-190.1)^2 + (158-190.1)^2 + (196-190.1)^2$
   $+ (195-190.1)^2 + (214-190.1)^2 + (199-190.1)^2 = 2440.88$

   $$s = \sqrt{\dfrac{2440.88}{7}} = 18.7 \text{ mm}$$

   d. $s^2 = 18.7^2 = 348.7 \text{ mm}^2$

   e.   The range is $214 - 158 = 56\,\text{mm}$

6. Yes. The frequencies increase to a maximum, and then they decrease. Also, the frequencies preceding the maximum are roughly a mirror image of those that follow the maximum.

7. No. Even though the sample is large, it is a voluntary response sample, so the responses cannot be considered to be representative of the population of the United States.

8. The vertical scale does not begin at 0, so the differences among different outcomes are exaggerated.

Chapter 4

Probability

Section 4-2, Basic Skills and Concepts

1. $P(A) = \dfrac{1}{10,000} = 0.0001$, $P(\bar{A}) = 1 - \dfrac{1}{10,000} = \dfrac{9999}{10,000} = 0.9999$

3. Part (c).

5. $5{:}2$, $\dfrac{7}{3}$, -0.9, $\dfrac{456}{123}$

7. $\dfrac{1}{5}$ or 0.2

9. Unlikely, neither unusually low nor unusually high

11. Unlikely, unusually low

13. $\dfrac{1}{4}$ or 0.25

15. $\dfrac{1}{2}$ or 0.5

17. $\dfrac{1}{5}$ or 0.2

19. 0

21. $\dfrac{6}{1000}$ or 0.006. The employer would suffer because it would be at a risk by hiring someone who uses drugs.

23. $\dfrac{50}{1000}$ or 0.05. This result is not close to the probability of 0.134 for a positive test result.

25. $\dfrac{879}{945}$ or 0.93. Yes, the technique appears to be effective.

27. $\dfrac{304}{300,000,000}$ or 0.00000101. No, the probability of being struck is much greater on an open golf course during a thunder storm. The golfer should seek shelter.

29. a. $\dfrac{1}{365}$ b. Yes

    c. He already knew d. 0

31. $\dfrac{10,427,000}{135,933,000}$ or 0.0767. No, a crash is not unlikely. Given that car crashes are so common, we should take precautions such as not driving after drinking and not using a cell phone or texting.

33. $\dfrac{8}{8+804} = 0.00985$. It is unlikely

35. $\dfrac{8}{492+8+306} = 0.00993$. Yes, it is unlikely. The middle seat lacks an outside view, easy access to the aisle, and a passenger in the middle seat has passengers on both sides instead of on one side only.

37. $\dfrac{3}{8}$ or 0.375

39. {bb, bg, gb, gg}; $\dfrac{1}{2}$ or 0.5

41. a. brown /brown, brown/blue, blue/brown, blue/blue

    b. $\dfrac{1}{4}$

    c. $\dfrac{3}{4}$

Section 4-2, Beyond the Basics

43. a. 999 : 1   b. 499 : 1

    c. The description is not accurate. The odds against winning are 999:1 and the odds in favor are 1:999, not 1:1000

45. a. $16   b. 8 : 1

    c. About 9.75 : 1, which becomes 39 : 4   d. $21.50

47. Relative risk: $\dfrac{\dfrac{26}{2103}}{\dfrac{22}{1671}} = 0.939$ Odds ratio: $\dfrac{\dfrac{\frac{26}{2103}}{1-\frac{26}{2103}}}{\dfrac{\frac{22}{1671}}{1-\frac{22}{1671}}} = 0.938$

The probability of a headache with Nasonex (0.0124) is slightly less than the probability of a headache with the placebo (0.0132), so Nasonex does not appear to pose a risk of headache.

49. $\dfrac{1}{4}$

Section 4-3, Basic Skills and Concepts

1. Based on the rule of the complements, the sum of $P(A)$ and its complement must always be 1, so the sum cannot be 0.5

3. Because it is possible to select someone who is male and a Republican, events M and R are not disjoint. Both events can occur at the same time when someone is randomly selected.

5. Disjoint

7. Not disjoint

9. Disjoint

11. Not disjoint

13. $1 - 0.47 = 0.53$

15. $P(\overline{D}) = 0.45$, where $P(\overline{D})$ is the probability of randomly selecting someone who does not choose a direct in-person encounter as the most fun way to flirt.

17. 1

19. $\dfrac{90 + 860 + 6}{1000} = 0.956$

21. $\dfrac{13}{28}$ or 0.464. That probability is not as high as it should be.

23. $\dfrac{16}{28}$ or 0.571

25. a. $\dfrac{11}{14} = 0.786$ or 78.6%   b. $\dfrac{2}{14} = 0.143$ or 14.3%

   c. The physicians given the labels with concentrations appear to have done much better. The results suggest that labels described as concentrations are much better than labels described as ratios.

Use the following table for Exercises 27–31

	Age						Total
	18–21	22–29	30–39	40–49	50–59	60 and over	
Responded	73	255	245	136	138	202	1049
Refused	11	20	33	16	27	49	156
Total	84	275	278	152	165	251	1205

27. $\dfrac{156}{1205} = 0.129$. Yes. A high refusal rate results in a sample that is not necessarily representative of the population, because those who refuse may well constitute a particular group with opinions different from others.

29. $\dfrac{1049}{1205} + \dfrac{84}{1205} - \dfrac{73}{1205} = \dfrac{1060}{1205} = 0.88$

31. $\dfrac{1049}{1205} + \dfrac{275 + 278}{1205} - \dfrac{255 + 245}{1205} = \dfrac{1102}{1205}$
   $= 0.915$

33. 300

	Positive Test Result	Negative Test Result	Total
Subject Used Marijuana	119	3	122
Subject Did not Use Marijuana	24	154	178
Total	143	157	300

35. $\dfrac{3 + 154 + 24}{300} = 0.603$

37. $\dfrac{27}{300} = 0.09$. With an error rate of 0.09 or 9%, the test does not appear to be highly accurate.

Section 4-3, Beyond the Basics

39. $\dfrac{3}{4}$ or 0.75

41. $P(A \text{ or } B) = P(A) + P(B) - 2P(A \text{ and } B)$

43. a. $1 - P(A) - P(B) + P(A \text{ and } B)$ b. $1 - P(A \text{ and } B)$

 c.   No

Section 4-4, Basic Skills and Concepts

1.   The probability that the second selected senator is a Democrat given that the first selected senator was a Republican.

3.   False. The events are dependent because the radio and air conditioner are both powered by the same electrical system. If you find that your car's radio does not work, there is a greater probability that the air conditioner will also not work.

5.   a.   The events are dependent   b. $\dfrac{1}{132}$ or 0.00758

7.   a.   Independent   b. $\dfrac{1}{12}$ or 0.0833

9.   a.   Independent   b. $\dfrac{5}{222} \cdot \dfrac{5}{222} = 0.000507$

11. a.   Dependent   b. $\dfrac{58}{100} \cdot \dfrac{1}{99} = 0.00586$

13. a. $\dfrac{90}{1000} \cdot \dfrac{90}{1000} = 0.0081$. Yes, it is unlikely   b. $\dfrac{90}{1000} \cdot \dfrac{89}{999} = 0.00802$. Yes, it is unlikely

15. a. $\dfrac{904}{1000} \cdot \dfrac{904}{1000} \cdot \dfrac{904}{1000} = 0.739$. No, it is not unlikely   b. $\dfrac{904}{1000} \cdot \dfrac{903}{999} \cdot \dfrac{902}{998} = 0.739$. No , it is not unlikely

17. $\dfrac{8330}{8834} \cdot \dfrac{8329}{8833} \cdot \dfrac{8328}{8832} = 0.838$. No, the entire batch consists of malfunctioning pacemakers.

19. a. $\dfrac{2}{100} = 0.02$

 b. $\dfrac{2}{100} \cdot \dfrac{2}{100} = 0.0004$

 c. $\dfrac{2}{100} \cdot \dfrac{2}{100} \cdot \dfrac{2}{100} = 0.000008$

 d.   By using one backup drive, the probability of failure is 0.02, and with three independent disk drives, the probability drops to 0.000008. By changing from one drive to three, the likelihood of failure drops from 1 chance in 50 to only 1 chance in 125,000, and that is a very substantial improvement in reliability. BACK UP YOUR DATA.

21. a. $\dfrac{1}{365}$ or 0.00274

   b. $\dfrac{1}{365}\cdot\dfrac{1}{365}=0.00000751$

   c. $\dfrac{1}{365}$ or 0.00274

23.

	Positive Test Result	Negative Test Result	Total
Subject used marijuana	True Positive 119	False Negative 3	122
Subject did not use marijuana	False Positive 24	True Negative 154	178
Total	143	157	300

$\dfrac{119}{300}\cdot\dfrac{118}{299}+\dfrac{154}{300}\cdot\dfrac{153}{299}+\dfrac{154}{300}\cdot\dfrac{119}{299}+\dfrac{119}{300}\cdot\dfrac{154}{299}=0.828$. No, it is not unlikely

25. $\dfrac{24}{300}\cdot\dfrac{23}{299}\cdot\dfrac{22}{298}=0.000454$. Yes, it is unlikely

27. a. $\dfrac{2518-252}{2518}=0.9$

   b. $\left(\dfrac{2518-252}{2518}\right)^{50}=0.00513$. Using the 5% guideline for cumbersome calculations

29. a. $\dfrac{162}{427}\cdot\dfrac{161}{426}=0.143$

   b. $\left(\dfrac{427-162}{427}\right)^{10}=0.00848$. Using the 5% guideline for cumbersome calculations

Section 4-4, Beyond the Basics

31. a. $0.99\cdot0.99+0.99\cdot0.01+0.01\cdot0.99=0.9999$

   b. $0.99\cdot0.99=0.9801$

   c.   The series arrangement provides better protection.

Section 4-5, Basic Skills and Concepts

1. a.   Answers vary, but 0.98 is a reasonable estimate.

   b.   Answers vary, but 0.999 is a reasonable estimate.

3.   The probability that the polygraph indicates lying given that the subject is actually telling the truth.

5.   At least one of the five children is a boy. $\dfrac{31}{32}$ or 0.969

7.   None of the digits is 0. $\left(\dfrac{9}{10}\right)^4=0.656$

9. $1-\left(\dfrac{4}{5}\right)^{10}=0.893$. The chance of passing is reasonably good

11. 0.5 or 50%

13. $1-\left(0.512\right)^{5}=0.965$

15. $1-\left(1-0.0423\right)^{3}=0.122$. Given that the three cars are in the same family, they are not randomly selected and there is a good chance that the family members have similar driving habits, so the probability might not be accurate.

17. $1-\left(1-0.67\right)^{4}=0.988$. It is very possible that the result is not valid because it is based on data from a voluntary response survey.

19. $\dfrac{90}{950}$ or 0.0947. This is the probability of the test making it appear that the subject uses drugs when the subject is not a drug user.

21. $\dfrac{6}{866}$ or 0.00693. This result is substantially different from the result found in Exercise 20. The probabilities P(subject uses drugs | negative test result) and P(negative test result | subject uses drugs) are not equal.

23. $\dfrac{44}{134}$ or 0.328

25. a. $\dfrac{1}{3}$ or 0.333   b. $\dfrac{5}{10}$ or 0.5

27. $\dfrac{10}{20}$ or 0.5

29. a. $1-\left(0.02\right)^{2}=0.9996$ b. $1-\left(0.02\right)^{3}=0.999992$

31. $1-\left(1-0.134\right)^{8}=0.684$. The probability is not low, so further testing of the individual samples will be necessary for about 68% of the combined samples.

Section 4-5, Beyond the Basics

33. a. $\dfrac{365}{365}\cdot\dfrac{364}{365}\cdot\dfrac{363}{365}\cdot\ldots\cdot\dfrac{341}{365}=0.431$ b. $1-0.431=0.569$

35. a. $\dfrac{0.8\cdot0.01}{0.8\cdot0.01+0.1\cdot0.99}=0.0748$ b.   0.8

     c.   The estimate of 75% is dramatically greater than the actual rate of 7.48%. They exhibited confusion of the inverse. A consequence is that they would unnecessarily alarm patients who are benign, and they might start treatments that are not necessary.

Section 4-6, Basic Skills and Concepts

1.   The symbol ! is the factorial symbol that represents the product of decreasing whole numbers, as in $4!=4\cdot3\cdot2\cdot1=24$. Four people can stand in line 24 different ways.

3. Because repetition is allowed, numbers are selected with replacement, so neither of the two permutation rules applies. The fundamental counting rule can be used to show that the number of possible outcomes is

$10 \cdot 10 \cdot 10 \cdot 10 = 10,000$, so the probability of winning is $\dfrac{1}{10,000}$.

5. $\dfrac{1}{10} \cdot \dfrac{1}{10} \cdot \dfrac{1}{10} \cdot \dfrac{1}{10} = \dfrac{1}{10,000}$

7. You will use Excel's PERMUT function to solve the problem. Click **FORMULAS** at the top of the screen. Select **Insert Function**. Select the **Statistical** category. Select the **PERMUT** function. Click **OK**. Enter the following values: Number = 9, Number chosen = 9. Click **OK**. The function returns 362880. The answer is

$\dfrac{1}{362880}$.

9. You will use Excel's COMBIN function to solve the problem. Click **FORMULAS** at the top of the screen. Select **Insert Function**. Select the **Math & Trig** category. Select the **COMBIN** function. Click **OK**. Enter the following values: Number = 27, Number chosen = 12. Click **OK**. The function returns 17,383,860. Because that number is so large, it is not practical to make a different CD for each possible combination.

11. You will use Excel's PERMUT function to solve the problem. Click **FORMULAS** at the top of the screen. Select **Insert Function**. Select the **Statistical** category. Select the **PERMUT** function. Click **OK**. Enter the following values: Number = 50, Number chosen = 4. Click **OK**. The function returns 5,527,200. The answer is

$\dfrac{1}{5527200}$. No, 5,527,200 is too many possibilities to list.

13. The formula is $\dfrac{11!}{4!4!2!}$. You will use Excel's FACT function to find the answer. The FACT function returns the factorial of a number. Enter this formula in an Excel worksheet:

    =FACT(11)/(FACT(4)*FACT(4)*FACT(2))

    Click **OK**. The answer is 34,650.

15. You will use Excel's COMBIN function to solve the problem. Click **FORMULAS** at the top of the screen. Select **Insert Function**. Select the **Math & Trig** category. Select the **COMBIN** function. Click **OK**. Enter the following values: Number = 44, Number chosen = 6. Click **OK**. The function returns 7,059,052. The answer is

$\dfrac{1}{7,059,052}$.

17. You will use Excel's PERMUT function to solve the problem. Click **FORMULAS** at the top of the screen. Select **Insert Function**. Select the **Statistical** category. Select the **PERMUT** function. Click **OK**. Enter the

following values: Number = 4, Number chosen = 4. Click **OK**. The function returns 24. The answer is $\dfrac{1}{24}$.

19. a. You will use Excel's COMBIN function to solve the problem. Click **FORMULAS** at the top of the screen. Select **Insert Function**. Select the **Math & Trig** category. Select the **COMBIN** function. Click **OK**. Enter the following values: Number = 41, Number chosen = 5. Click **OK**. The function returns 749,398. The

answer is $\dfrac{1}{749,398}$.

    b. $\dfrac{1}{10^4} = \dfrac{1}{10,000}$

    c. $10,000

21. a. You will use Excel's PERMUT function to solve the problem. Click **FORMULAS** at the top of the screen. Select **Insert Function**. Select the **Statistical** category. Select the **PERMUT** function. Click **OK**. Enter the

following values: Number = 12, Number chosen = 4. Click **OK**. The function returns 11,880. Officers can be appointed in 11,880 ways.

   b. You will use Excel's COMBIN function to solve the problem. Click **FORMULAS** at the top of the screen. Select **Insert Function**. Select the **Math & Trig** category. Select the **COMBIN** function. Click **OK**. Enter the following values: Number = 12, Number chosen = 4. Click **OK**. The function returns 495. The committee can be appointed in 495 different ways.

   c. $\dfrac{1}{495}$

23. You will use Excel's PERMUTATIONA function to solve the problem. Click **FORMULAS** at the top of the screen. Select **Insert Function**. Select the **All** category. Select the **PERMUTATIONA** function. Click **OK**. Enter the following values: Number = 50, Number chosen = 3. Click **OK**. The function returns 125,000. The fundamental counting rule can be used. The different possible codes are ordered sequences of numbers, not combinations, so the name of "combination lock" is not appropriate. Given that "fundamental counting rule lock" is a bit awkward, a better name would be something like "number lock".

25. You will use Excel's PERMUT function to solve the problem. Click **FORMULAS** at the top of the screen. Select **Insert Function**. Select the **Statistical** category. Select the **PERMUT** function. Click **OK**. Enter the following values: Number = 5, Number chosen = 5. Click **OK**. The answer is 120 ways. The correct unscrambling is *AMITY*. The probability of getting that result by randomly selecting one arrangement is $\dfrac{1}{120}$.

27. You will use Excel's COMBIN function to obtain four results that will then be added together. Click **FORMULAS** at the top of the screen. Select **Insert Function**. Select the **Math & Trig** category. Select the **COMBIN** function. Click **OK**.

   1st: Number = 5, Number chosen = 2. The function returns 10.

   2nd: Number = 5, Number chosen = 3. The function returns 10.

   3rd: Number = 5, Number chosen = 4. The function returns 5.

   4th: Number = 5, Number chosen = 5. The function returns 1.

   The answer is found by adding these four results together, $10 + 10 + 5 + 1 = 26$.

29. You will use Excel's COMBIN function to solve the problem. Click **FORMULAS** at the top of the screen. Select **Insert Function**. Select the **Math & Trig** category. Select the **COMBIN** function. Click **OK**. Enter the following values: Number = 5, Number chosen = 2. Click **OK**. The function returns 120. 120 different tests are required for every possible pairing of two wires.

31. $4 \cdot 4 \cdot 4 = 64$

33. $\dfrac{1}{{}_{59}C_5 \cdot {}_{39}C_1} = \dfrac{1}{195,249,054}$

35. $\dfrac{2}{{}_{10}C_5} = \dfrac{2}{252}$. Yes, if everyone treated is of one gender while everyone in the placebo group is of the opposite gender, you would not know if different reactions are due to the treatment or gender.

Section 4-6, Beyond the Basics

37. $26 + 26 \cdot 36 + 26 \cdot 36^2 + 26 \cdot 36^3 + 26 \cdot 36^4 + 26 \cdot 36^5 + 26 \cdot 36^6 + 26 \cdot 36^7 = 2,095,681,645,538$ (about 2 trillion).

39. 12 ways: {25p, 1n 20p, 2n 15p, 3n 10p, 4n 5p, 5n, 1d 15p, 1d 1n 10p, 1d 2n 5p, 1d 3n, 2d 5p,2d 1n} (Note: 25p represents 25 pennies, etc.)

Chapter Quick Quiz

1. 0 (not an option)

2. $\dfrac{10-3}{10}=0.7$

3. 1 (all days contain the letter y)

4. $0.2\cdot0.2=0.04$

5. Answers vary, but an answer such as 0.01 or lower is reasonable

6. $\dfrac{288+224}{201+126+288+224}=\dfrac{512}{839}=0.61$

7. $\dfrac{224+288+201}{839}=\dfrac{713}{839}=0.85$

8. $\dfrac{126}{839}=0.15$

9. $\dfrac{126}{839}\cdot\dfrac{125}{839}=0.0224$

10. $\dfrac{126}{126+224}=\dfrac{126}{350}=0.36$

Review Exercises

1. $\dfrac{392+58}{1028}=0.438$

2. $\dfrac{392}{392+564}=0.41$

3. $\dfrac{58}{58+14}=0.806$

4. It appears that you have a substantially better chance of avoiding prison if you enter a guilty plea.

5. $\dfrac{392+58}{1028}+\dfrac{392+564}{1028}-\dfrac{392}{1028}=0.986$

6. $\dfrac{450}{1028}\cdot\dfrac{449}{1027}=0.191$

7. $\dfrac{72}{1028}\cdot\dfrac{71}{1027}=0.00484$

8. $\dfrac{72}{1028}+\dfrac{578}{1028}-\dfrac{14}{1028}=0.619$

9. $\dfrac{392}{1028}=0.381$

10. $\dfrac{14}{1028}=0.0136$

11. Answers vary, but DuPont data show that about 8% of cars are red, so any estimate between 0.01 and 0.2 would be reasonable.

12. a. $1-0.35=.65$

    b. $(0.35)^4=0.015$

    c. Yes, because the probability is so small

13. a. $\dfrac{1}{365}$

    b. $\dfrac{31}{365}$

    c. Answers vary, but it is probably small, such as 0.02

    d. Yes

14. $1-\left(1-\dfrac{213}{100,000}\right)^{10}=0.0211$. No

15. You will use Excel's COMBIN function to solve the problem. Click **FORMULAS** at the top of the screen. Select **Insert Function**. Select the **Math & Trig** category. Select the **COMBIN** function. Click **OK**. Enter the following values: Number = 42, Number chosen = 6. Click **OK**. The function returns 5,245,786. The answer is $\dfrac{1}{5,245,786}$.

16. You will use Excel's COMBIN function to solve the problem. Click **FORMULAS** at the top of the screen. Select **Insert Function**. Select the **Math & Trig** category. Select the **COMBIN** function. Click **OK**. Enter the following values: Number = 39, Number chosen = 5. Click **OK**. The function returns 575,757. The answer is

$$\frac{1}{575,757}.$$

17 You will use Excel's PERMUTATIONA function to solve the problem. Click **FORMULAS** at the top of the screen. Select **Insert Function**. Select the **All** category. Select the **PERMUTATIONA** function. Click **OK**. Enter the following values: Number = 10, Number chosen = 3. Click **OK**. The function returns 1,000. The

probability is $\dfrac{1}{1,000}$.

18. You will use Excel's PERMUT function to solve the problem. Click **FORMULAS** at the top of the screen. Select **Insert Function**. Select the **All** category. Select the **PERMUTATIONA** function. Click **OK**. Enter the following values: Number = 12, Number chosen = 3. Click **OK**. The function returns 1,320. 1,320 different

trifecta bets are possible. The probability of winning is $\dfrac{1}{1,320}$.

Cumulative Review Exercises

1. a. The mean of –8.9 years is not close to the value of 0 years that would be expected with no gender discrepancy.

    b. The median of –13.5 years is not close to the value of 0 years that would be expected with no gender discrepancy.

    c. $s = \sqrt{\dfrac{\left(-20-(-8.9)\right)^2 + \left(-15-(-8.9)\right)^2 + \ldots + \left(-15-(-8.9)\right)^2}{11}} = 10.6$ years.

    d. $s^2 = (10.6)^2 = 113.2$ years^2

    e. $Q_1 = -15$ years

    f. $Q_3 = -5$ years

    g. The boxplot suggests that the data have a distribution that is skewed.

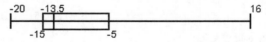

2. a. $z = \dfrac{100-77.5}{11.6} = 1.94$. No, the pulse rate of 100 beats per minute is within 2 standard deviations away from the mean, so it is not unusual.

    b. $z = \dfrac{50-77.5}{11.6} = -2.37$. Yes, the pulse rate of 50 beats per minutes is more than 2 standard deviations away from the mean so it is unusual.

    c. Yes, because the probability of $\dfrac{1}{256}$ (or 0.0039) is so small.

    d. No, because the probability of $\dfrac{1}{8}$ (or 0.125) is not very small.

3. a. $\dfrac{2346}{5100} = .46 = 46\%$

    b. $0.46 = 46\%$

c.   Stratified sample

4.   The graph is misleading because the vertical scale does not start at 0. The vertical scale starts at the frequency of 500 instead of 0, so the difference between the two response rates is exaggerated. The graph incorrectly makes it appear that "no" responses occurred 60 times more often than the number of "yes" responses, but comparisons of the actual frequencies shows that the "no" responses occurred about four times more often than the number of "yes" responses.

5.   a.   A convenience sample

b.   If the students at the college are mostly from a surrounding region that includes a large proportion of one ethnic group, the results will not reflect the general population of the United States.

c. $0.35 + 0.4 = 0.75$

d. $1 - (0.6)^2 = 0.64$

6.   The straight-line pattern of the points suggests that there is a correlation between chest size and weight.

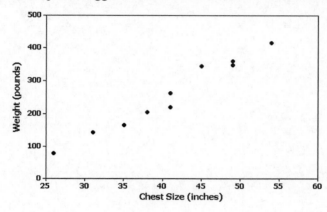

7.   a. $\dfrac{1}{_{39}C_5} = \dfrac{1}{575{,}757}$ b. $\dfrac{1}{19}$

c. $\dfrac{1}{_{39}C_5 \cdot {_{19}C_1}} = \dfrac{1}{10{,}939{,}383}$

Chapter 5

Discrete Probability Distributions

Section 5-2, Basic Skills and Concepts

1. The random variable is x, which is the number of girls in three births. The possible values of x are 0, 1, 2, and 3. The values of the random variable x are numerical.

3. Table 5-7 does describe a probability distribution because the three requirements are satisfied. First, the variable x is a numerical random variable and its values are associated with probabilities. Second, $\Sigma P(x) = 0.125 + 0.375 + 0.375 + 0.125 = 1$ as required. Third, each of the probabilities is between 0 and 1 inclusive, as required.

5. a. Continuous random variable

    b. Discrete random variable

    c. Not a random variable

    d. Discrete random variable

    e. Continuous random variable

    f. Discrete random variable

7. You will use formulas in an Excel worksheet to carry out the calculations.

    Step one: One of the three requirements of a probability distribution is that $\sum P(x) = 1$. In addition to checking whether the other two requirements are met, you need to see if $\sum P(x) = 1$. Enter the labels in row 1 of an Excel worksheet as shown below. Next, enter the x values in column B and the P(x) values in column C. Use a formula to calculate the sum of P(x). Because the other two requirements are met and $\sum P(x) = 1$, a probability distribution is given.

	A	B	C
1		x	P(x)
2		0	0.0625
3		1	0.25
4		2	0.375
5		3	0.25
6		4	0.0625
7	Sum		1

    Step two: Because a probability distribution is given, you will find its mean and then its standard deviation. Enter the labels in cells D1 and E1 as displayed below. Note that these labels are Formulas 5-1 and 5-2 shown on page 200 of your textbook. Next, input formulas in the Excel worksheet to calculate the $x*P(x)$ products in column D and then use a formula to calculate the sum of the products.

	A	B	C	D	E
1		x	P(x)	x*P(x)	(x-mu)^2*P(x)
2		0	0.0625	0	
3		1	0.25	0.25	
4		2	0.375	0.75	
5		3	0.25	0.75	
6		4	0.0625	0.25	
7	Sum		1	2	

    Mean = $\mu = \sum x \cdot P(x) = 2$.

Step three: Using the value that you obtained for the mean, you will calculate the variance and the standard deviation. Input formulas in column E of the Excel worksheet to carry out the calculations indicated. Then use a formula to calculate the sum of the column E values.

$$\text{Variance} = \sigma^2 = \sum \left[(x - \mu)^2 \cdot P(x) \right] = 1$$

Take the square root of the variance to obtain the standard deviation. Standard deviation = $\sqrt{\sigma^2} = \sqrt{1} = 1$.

	A	B	C	D	E
1		x	P(x)	x*P(x)	(x-mu)^2*P(x)
2		0	0.0625	0	0.25
3		1	0.25	0.25	0.25
4		2	0.375	0.75	0
5		3	0.25	0.75	0.25
6		4	0.0625	0.25	0.25
7	Sum		1	2	1
8				Standard Deviation	1

The formulas are displayed in the Excel worksheet below. Note that the formulas use cell addresses rather than numerical values.

	A	B	C	D	E
1		x	P(x)	x*P(x)	(x-mu)^2*P(x)
2		0	0.0625	=B2*C2	=(B2-D7)^2*C2
3		1	0.25	=B3*C3	=(B3-D7)^2*C3
4		2	0.375	=B4*C4	=(B4-D7)^2*C4
5		3	0.25	=B5*C5	=(B5-D7)^2*C5
6		4	0.0625	=B6*C6	=(B6-D7)^2*C6
7	Sum		=SUM(C2:C6)	=SUM(D2:D6)	=SUM(E2:E6)
8				Standard Deviation	=SQRT(E7)

9. Follow the procedure presented in Exercise 7. Not a probability distribution because the sum of the probabilities is 0.601, which is not 1 as required. Also, Ted clearly needs a new approach.

11. Follow the procedure presented in Exercise 7. Probability distribution with $\mu = 2.2$ and $\sigma = 1.00$. (The sum of the probabilities is 0.999, but that is due to rounding errors.)

	A	B	C	D	E
1		x	P(x)	x*P(x)	(x-mu)^2*P(x)
2		0	0.041	0	0.198259641
3		1	0.2	0.2	0.2875202
4		2	0.367	0.734	0.014533567
5		3	0.299	0.897	0.191838699
6		4	0.092	0.368	0.298411292
7	Sum		0.999	2.199	0.990563399
8				Standard Deviation	0.995270515

13. Follow the procedure presented in Exercise 7. Not a probability distribution because the responses are not values of a numerical random variable. Also, sum of the probabilities is 1.18 instead of 1 as required.

	A	B	C
1	x		P(x)
2		Family/partner	0.77
3		Friends	0.15
4		Work/studies	0.08
5		Leisure	0.08
6		Music	0.06
7		Sports	0.04
8	Sum		1.18

15. Follow the procedure presented in Exercise 7. $\mu = 5$ and $\sigma = 1.6$.

	A	B	C	D	E
1		x	P(x)	x*P(x)	(x-mu)^2*P(x)
2		0	0.001	0	0.025
3		1	0.01	0.01	0.16
4		2	0.044	0.088	0.396
5		3	0.117	0.351	0.468
6		4	0.205	0.82	0.205
7		5	0.246	1.23	0
8		6	0.205	1.23	0.205
9		7	0.117	0.819	0.468
10		8	0.044	0.352	0.396
11		9	0.01	0.09	0.16
12		10	0.001	0.01	0.025
13	Sum		1	5	2.508
14				Standard Deviation	1.583666632

17. a. $P(X = 8) = 0.044$

b. $P(X \geq 8) = 0.044 + 0.01 + 0.001 = 0.055$

c.   The result from part (b)

d.   No, because the probability of 8 or more girls is 0.055, which is not very low (less than or equal to 0.05)

19. Follow the procedure presented in Exercise 7. $\mu = 0.9$ and $\sigma = 0.9$.

	A	B	C	D	E
1		x	P(x)	x*P(x)	(x-mu)^2*P(x)
2		0	0.377	0	0.301311972
3		1	0.399	0.399	0.004483164
4		2	0.176	0.352	0.215289536
5		3	0.041	0.123	0.181844676
6		4	0.005	0.02	0.04823618
7		5	0	0	0
8		6	0	0	0
9	Sum		0.998	0.894	0.751165528
10				Standard Deviation	0.86669806

21. a. $P(X = 3) = 0.041$

b. $P(X \geq 3) = 0.041 + 0.005 + 0 + 0 = 0.046$

c.   The probability from part (b)

d.   Yes, because the probability of three or more failures is 0.046 which is very low (less than or equal to 0.05).

Section 5-2, Beyond the Basics

23. a.   You will use Excel's PERMUTATIONA function to solve the problem. Click **Formulas** at the top of the screen. Select **Insert Function**. Select the **All** category. Select the **PERMUTATIONA** function. Click **OK**. Enter the following values: Number = 10, Number chosen = 3. Click **OK**. The function returns 1,000.

   b. $\dfrac{1}{1,000}$

   c. $\$500 - \$1 = \$499$

   d. $-\$1 \cdot 1 + \$500 \cdot \dfrac{1}{1000} = -\$0.50 = -50 \text{ cents}$

   e.   The $1 bet on the pass line in craps is better because its expected value of -1.4 cents is much greater than the expected value of -50 cents for the Texas Pick 3 lottery.

25. a. $-\$0.26 + \$30 \cdot \dfrac{5}{38} - \$5 \cdot \dfrac{33}{38} = -0.39$

   b.   The bet on the number 27 is better because its expected value of -26 cents is greater than the expected value of -39 cents for the other bet.

Section 5-3, Basic Skills and Concepts

1.   The given calculation assumes that the first two adults include Wal-Mart and the last three adults do not include Wal-Mart, but there are other arrangements consisting of two adults who include Wal-Mart and three who do not. The probabilities corresponding to those other arrangements should also be included in the result.

3.   Because the 30 selections are made without replacement, they are dependent, not independent. Based on the 5% guideline for cumbersome calculations, the 30 selections can be treated as being independent. (The 30 selections constitute 3% of the population of 1000 responses, and 3% is not more than 5% of the population.) The probability can be found by using the binomial probability formula.

5.   Not binomial. Each of the weights has more than two possible outcomes.

7.   Binomial

9.   Not binomial. Because the senators are selected without replacement, the selections are not independent. (The 5% guideline for cumbersome calculations cannot be applied because the 40 selected senators constitute 40% of the population of 100 senators, and that exceeds 5%.)

11.   Binomial. Although the events are not independent, they can be treated as being independent by applying the 5% guideline. The sample size of 380 is no more than 5% of the population of all smartphone users.

13. a. $\dfrac{4}{5} \cdot \dfrac{4}{5} \cdot \dfrac{1}{5} = 0.128$

   b. $\{WWC, WCW, CWW\}$; 0.128 for each

   c. $0.128 \cdot 3 = 0.384$

15. $_5C_3 \cdot 0.2^3 \cdot 0.8^2 = 0.051$

17. $_5C_3 \cdot 0.2^3 \cdot 0.8^2 + {}_5C_4 \cdot 0.2^4 \cdot 0.8^1 + {}_5C_5 \cdot 0.2^5 \cdot 0.8^0 = 0.057$

19. $_5C_5 \cdot 0.2^0 \cdot 0.8^5 = 0.328$

21. $_8C_3 \cdot 0.45^3 \cdot 0.55^5 = 0.257$

You can use Excel's BINOM.DIST function to solve the problem. Click **Formulas** at the top of the screen. Select **Insert Function**. Select the **Statistical** category. Select the **BINOM.DIST** function. Click **OK**. Enter the following values: Number of successes = 3, Trials = 8, Probability of success = 0.45, Cumulative = FALSE. Click **OK**. The function returns 0.2568.

23. $_{20}C_{16} \cdot 0.45^{16} \cdot 0.55^{4} = 0.00125$

You can use Excel's BINOM.DIST function to solve the problem. Click **Formulas** at the top of the screen. Select **Insert Function**. Select the **Statistical** category. Select the **BINOM.DIST** function. Click **OK**. Enter the following values: Number of successes = 16, Trials = 20, Probability of success = 0.45, Cumulative = FALSE. Click **OK**. The function returns 0.00125.

25. $P(X \geq 2) = 0.033 + 0.132 + 0.297 + 0.356 + 0.178 = 0.996$; yes

27. $P(X \leq 2) = 0.000 + 0.004 + 0.033 = 0.037$; yes, because the probability of 2 or fewer peas with green pods is small (less than or equal to 0.05).

29. a.   You will use Excel's BINOM.DIST function to solve the problem. Click **Formulas** at the top of the screen. Select **Insert Function**. Select the **Statistical** category. Select the **BINOM.DIST** function. Click **OK**. Enter the following values: Number of successes = 5, Trials = 6, Probability of success = 0.2, Cumulative = FALSE. Click **OK**. The function returns 0.001536.

   b.   Select the **BINOM.DIST** function. Click **OK**. Enter the following values: Number of successes = 6, Trials = 6, Probability of success = 0.2, Cumulative = FALSE. Click OK. The function returns 6.4E-05.

   c.   0.001536 + 6.4E-05 = 0.001536.

   d.   Yes, the small probability from part (c) suggests that 5 is an unusually high number.

31. a.   You will use Excel's BINOM.DIST function to solve the problem. Click **Formulas** at the top of the screen. Select **Insert Function**. Select the **Statistical** category. Select the **BINOM.DIST** function. Click **OK**. Enter the following values: Number of successes = 0, Trials = 5, Probability of success = 0.2, Cumulative = FALSE. Click **OK**. The function returns 0.3277.

   b.   Select the **BINOM.DIST** function. Click **OK**. Enter the following values: Number of successes = 1, Trials = 5, Probability of success = 0.2, Cumulative = FALSE. Click OK. The function returns 0.4096.

   c.   From part (a) above, P(x=0) = 0.3277. From part (b) above, P(x=1) = 0.4096. 0.3277 + 0.4096 = 0.7373.

   d.   No, the probability from part (c) is not small, so 1 is not an unusually low number.

33. You will use Excel's BINOM.DIST function to solve the problem. Click **Formulas** at the top of the screen. Select **Insert Function**. Select the **Statistical** category. Select the **BINOM.DIST** function. Click **OK**. Enter the following values: Number of successes = 12, Trials = 20, Probability of success = 0.48, Cumulative = FALSE. Click **OK**. The function returns 0.1007. No, because the probability of exactly 12 is 0.1007, the probability of 12 or more is greater than 0.1007, so the probability of getting 12 or more is not very small, so 12 is not unusually high.

35. You will use Excel's BINOM.DIST function to solve the problem. Click **Formulas** at the top of the screen. Select **Insert Function**. Select the **Statistical** category. Select the **BINOM.DIST** function. Click **OK**. Enter the following values: Number of successes = 10, Trials = 12, Probability of success = 0.805, Cumulative = FALSE. Click **OK**. The function returns 0.2868. No, because the flights all originate from New York, they are not randomly selected flights, so the 80.5% on-time rate might not apply.

37. a.   You will use Excel's BINOM.DIST function to solve the problem. Click **Formulas** at the top of the screen. Select **Insert Function**. Select the **Statistical** category. Select the **BINOM.DIST** function. Click **OK**. Enter the following values: Number of successes = 0, Trials = 12, Probability of success = 0.45, Cumulative = FALSE. Click **OK**. The function returns 0.000766.

   b.   1 – 0.000766 = .999234

   c.   Select the **BINOM.DIST** function. Click **OK**. Enter the following values: Number of successes = 1, Trials = 12, Probability of success = 0.45, Cumulative = TRUE. Click OK. The function returns 0.00829.

d.   Yes, the very low probability of 0.00829 would suggest that the 45 share value is wrong.

39. a.   You will use Excel's BINOM.DIST function to solve the problem. Click **Formulas** at the top of the screen. Select **Insert Function**. Select the **Statistical** category. Select the **BINOM.DIST** function. Click **OK**. Enter the following values: Number of successes = 13, Trials = 14, Probability of success = 0.5, Cumulative = FALSE. Click **OK**. The function returns 0.000854.

   b.   Select the **BINOM.DIST** function. Click **OK**. Enter the following values: Number of successes = 14, Trials = 14, Probability of success = 0.5, Cumulative = FALSE. Click **OK**. The function returns 6.10352E-05.

   c.   From part (a) above, P(x=13) = 0.000854. From part (b) above, P(x=14) = 0.0000610. 0.000854 + 0.0000610 = 0.000915.

   d.   Yes. The probability of getting 13 girls or a result that is more extreme is 0.000915, so chance does not appear to be a reasonable explanation for the result of 13 girls. Because 13 is an unusually high number of girls, it appears that the probability of a girl is higher with the XSORT method, and it appears that the XSORT method is effective.

41.   First find the probability that 0 of 24 test positive. Then subtract that result from 1, which will give you the probability that 1 or more will test positive. Click **Formulas** at the top of the screen. Select **Insert Function**. Select the **Statistical** category. Select the **BINOM.DIST** function. Click **OK**. Enter the following values: Number of successes = 0, Trials = 24, Probability of success = 0.006, Cumulative = FALSE. Click **OK**. The function returns 0.8655. 1 − 0.8655 = 0.1345. It is not unlikely for such a combined sample to test positive.

43.   You will use Excel's BINOM.DIST function to solve the problem. Click **Formulas** at the top of the screen. Select **Insert Function**. Select the **Statistical** category. Select the **BINOM.DIST** function. Click **OK**. Enter the following values: Number of successes = 1, Trials = 40, Probability of success = 0.03, Cumulative = TRUE. Click **OK**. The function returns 0.6615. The probability shows that about 2/3 of all shipments will be accepted. With about 1/3 of the shipments rejected, the supplier would be wise to improve quality.

Section 5-3, Beyond the Basics

45. $P(X = 5) = 0.06(1 - 0.06)^4 = 0.0468$

47. a. $P(4) = \dfrac{6!}{(6-4)!4!} \cdot \dfrac{43!}{(43-6+4)!(6-4)!} \div \dfrac{(6+43)!}{(6+43-6)!6!} = 0.000969$

   b. $P(6) = \dfrac{6!}{(6-6)!6!} \cdot \dfrac{43!}{(43-6+6)!(6-6)!} \div \dfrac{(6+43)!}{(6+43-6)!6!} = 0.0000000715$

   c. $P(0) = \dfrac{6!}{(6-0)!0!} \cdot \dfrac{43!}{(43-6+0)!(6-0)!} \div \dfrac{(6+43)!}{(6+43-6)!6!} = 0.436$

Section 5-4, Basic Skills and Concepts

1. $n = 270, p = 0.07, q = 0.93$

3.   Variance is $150 \cdot 0.933 \cdot 0.067 = 9.4$ executives^2

5. $\mu = np = 60 \cdot 0.2 = 12$ correct guesses and $\sigma = \sqrt{np(1-p)} = \sqrt{60 \cdot 0.2 \cdot 0.8} = 3.1$ correct guesses. Minimum: $12 - 2(3.1) = 5.8$ correct guesses, maximum: $12 + 2(3.1) = 18.2$ correct guesses.

7. $\mu = np = 1013 \cdot 0.66 = 668.6$ worriers and $\sigma = \sqrt{np(1-p)} = \sqrt{1013 \cdot 0.66 \cdot 0.34} = 15.1$ worriers. Minimum: $668.6 - 2(15.1) = 638.4$ worriers, maximum: $668.6 + 2(15.1) = 698.8$ worriers

9. a. $\mu = np = 291 \cdot 0.5 = 145.5$ boys and $\sigma = \sqrt{np(1-p)} = \sqrt{291 \cdot 0.5 \cdot 0.5} = 8.5$ boys

   b.   Yes. Using the range rule of thumb, the minimum value is $145.5 - 2(8.5) = 128.5$ boys and the maximum value is $145.5 + 2(8.5) = 162.5$ boys. Because 239 boys is above the range of usual values, it is unusually high. Because 239 boys is unusually high, it does appear that the YSORT method of gender selection is effective.

11. a. $\mu = np = 100 \cdot 0.20 = 20$ and $\sigma = \sqrt{np(1-p)} = \sqrt{100 \cdot 0.2 \cdot 0.8} = 4$

   b.   No, because 25 orange M&Ms is within the range of usual values of $20 - 2(4) = 12$ and $20 + 2(4) = 28$. The claimed rate of 20% does not necessarily appear to be wrong, because that rate will usually result in 12 to 28 orange M&Ms (among 100), and the observed number of orange M&Ms is within that range.

13. a. $\mu = np = 420,095 \cdot 0.00034 = 142.8$ and $\sigma = \sqrt{np(1-p)} = \sqrt{420,095 \cdot 0.00034 \cdot 0.999666} = 11.9$

   b.   No, 135 is not unusually low or high because it is within the range of usual values $142.8 - 2(11.9) = 119$ and $142.8 + 2(11.9) = 166.6$

   c.   Based on the given results, cell phones do not pose a health hazard that increases the likelihood of cancer of the brain or nervous system.

15. a. $\mu = np = 2600 \cdot 0.06 = 156$ and $\sigma = \sqrt{np(1-p)} = \sqrt{2600 \cdot 0.06 \cdot 0.94} = 12.1$

   b.   The minimum usual frequency is $156 - 2(12.1) = 131.8$ and the maximum is $156 + 2(12.1) = 180.2$. The occurrence of r 178 times is not unusually low or high because it is within the range of usual values.

17. a. $\mu = np = 370 \cdot 0.2 = 74$ and $\sigma = \sqrt{np(1-p)} = \sqrt{370 \cdot 0.2 \cdot 0.8} = 7.7$

   b.   The minimum usual number is $74 - 2(7.7) = 58.6$ and the maximum is $74 + 2(7.7) = 89.4$. The value of 90 is unusually high because it is above the range of usual values.

19. a. $\mu = np = 30 \cdot \dfrac{1}{365} = 0.0821918$ and $\sigma = \sqrt{np(1-p)} = \sqrt{30 \cdot \dfrac{1}{365} \cdot \dfrac{364}{365}} = 0.2862981$

   b.   The minimum usual value is $0.0821918 - 2(0.2862981) = -0.4904044$ and the maximum is $0.0821918 + 2(0.2862981) = 0.654788$. The result of 2 students born on the 4th of July would be unusually high, because 2 is above the range of usual values.

Section 5-4, Beyond the Basics

21. From the range of usual values we get $\mu = 60$ and $\sigma = 6$. Using the formulas for the mean and the standard deviation we get $p = 1 - \dfrac{\sigma^2}{\mu} = 0.4$ which leads to $n = \dfrac{\mu}{p} = 150$ and $q = 0.6$

23. The probability of selecting a girl out of 40 is given by $P(X = n) = \dfrac{_{10}C_n \cdot {}_{30}C_{12-n}}{_{40}C_{12}}$ the table on the next page lists the probabilities of selecting the number of girls from 0 to 10.

Number of girls ($X = n$)	Probability $P(X = n)$
0	0.0154815
1	0.0977782
2	0.2420012
3	0.3073032
4	0.2200011
5	0.0918266
6	0.0223190
7	0.0030608
8	0.0002207
9	0.0000073
10	0.0000001

The mean is $\mu = \sum [x \cdot P(x)]$

$\mu = 0 \cdot 0.0154815 + 1 \cdot 0.0977782 + 2 \cdot 0.2420012 + 3 \cdot 0.3073032 + 4 \cdot 0.2200011 + 5 \cdot 0.0918266 +$
$\quad 6 \cdot 0.0223190 + 7 \cdot 0.0030608 + 8 \cdot 0.0002207 + 9 \cdot 0.0000073 + 10 \cdot 0.0000001 = 3$

The standard deviation is $\sigma = \sqrt{\sum [x^2 \cdot P(x)] - \mu^2}$

$\sigma = \sqrt{\begin{array}{l} 0^2 \cdot 0.0154815 + 1^2 \cdot 0.0977782 + 2^2 \cdot 0.2420012 + 3^2 \cdot 0.3073032 + 4^2 \cdot 0.2200011 + 5^2 \cdot 0.0918266 + \\ 6^2 \cdot 0.0223190 + 7^2 \cdot 0.0030608 + 8^2 \cdot 0.0002207 + 9^2 \cdot 0.0000073 + 10^2 \cdot 0.0000001 - 3^3 \end{array}}$

$\quad = 1.3$

Section 5-5, Basic Skills and Concepts

1. $\mu = \dfrac{535}{576} = 0.929$, which is the mean number of hits per region. $x = 2$, because we want the probability that a randomly selected region had exactly 2 hits, and $e = 2.71828$ which is a constant used in all applications of Formula $5 - 9$.

3. With $n = 50$, the first requirement of $n \geq 100$ is not satisfied. With $n = 50$ and $p = 0.001$ the second requirement of $np \leq 10$ is satisfied. Because both requirements are not satisfied, we should not use the Poisson distribution as an approximation to the binomial.

5. You will use Excel's POISSON.DIST function to solve the problem. Click **Formulas** at the top of the screen. Select **Insert Function**. Select the **Statistical** category. Select the **POISSON.DIST** function. Click **OK**. Enter the following values: x = 0, Mean = 8.5, Cumulative = FALSE. Click **OK**. The function returns 0.000203. Yes, it is unlikely.

7. You will use Excel's POISSON.DIST function to solve the problem. Click **Formulas** at the top of the screen. Select **Insert Function**. Select the **Statistical** category. Select the **POISSON.DIST** function. Click **OK**. Enter the following values: x = 10, Mean = 8.5, Cumulative = FALSE. Click **OK**. The function returns 0.1104. No, it is not unlikely.

9. a. $\mu = \dfrac{268}{41} = 6.5$

   b. First find the probability of no earthquakes that measure 6.0 or higher on the Richter scale. Then subtract that result from 1, which will give you the probability that there will be at least one. Click **Formulas** at the top of the screen. Select **Insert Function**. Select the **Statistical** category. Select the **POISSON.DIST** function. Click **OK**. Enter the following values: x = 0, Mean = 6.5, Cumulative = FALSE. Click **OK**. The function returns 0.0015. $P(x \geq 1) = 1 - 0.0015 = 0.9985$.

   c. Yes. Based on the result in part (b), we are quite sure (with probability 0.999) that there is at least one earthquake measuring 6.0 or higher on the Richter scale, so there is a very low probability (0.001) that there will be no such earthquake in a year.

11. a. $\mu = \dfrac{22713}{365} = 62.2$

   b. You will use Excel's POISSON.DIST function to solve the problem. Click **Formulas** at the top of the screen. Select **Insert Function**. Select the **Statistical** category. Select the **POISSON.DIST** function. Click **OK**. Enter the following values: x = 50, Mean = 62.2, Cumulative = FALSE. Click **OK**. The function returns 0.0156.

13. a. You will use Excel's POISSON.DIST function to solve the problem. Click **Formulas** at the top of the screen. Select **Insert Function**. Select the **Statistical** category. Select the **POISSON.DIST** function. Click **OK**. Enter the following values: x = 2, Mean = 0.929, Cumulative = FALSE. Click **OK**. The function returns 0.1704.

   b. $(576)(0.1704) = 98.15$

   c. The expected number of regions with 2 hits is close to 93, which is the actual number of regions with 2 hits.

15. a. You will use Excel's POISSON.DIST function to solve the problem. Click **Formulas** at the top of the screen. Select **Insert Function**. Select the **Statistical** category. Select the **POISSON.DIST** function. Click **OK**. Enter the following values: x = 26, Mean = 30.4, Cumulative = FALSE. Click **OK**. The function returns 0.05579. The expected number = $(34)(0.05579) = 1.897$. The expected number of cookies is 1.9, and that is very close to the actual number of cookies with 26 chocolate chips, which is 2.

   b. Select the **POISSON.DIST** function. Click **OK**. Enter the following values: x = 30, Mean = 30.4, Cumulative = FALSE. Click **OK**. The function returns 0.07244. The expected number = $(34)(0.07244) = 2.463$. The expected number of cookies is 2.5, and that is very different from the actual number of cookies with 30 chocolate chips which is 6.

Section 5-5, Beyond the Basics

17. a. No. With $n = 12$ and $p = \frac{1}{6}$ the requirement of $n \geq 100$ is not satisfied, so the Poisson distribution is not a good approximation to the binomial distribution.

   b. You will use Excel's POISSON.DIST and BINOMIN.DIST functions to answer the question. Click **Formulas** at the top of the screen. Select **Insert Function**. Select the **Statistical** category. Select the **POISSON.DIST** function. Click **OK**. Enter the following values: x = 3, Mean = 2, Cumulative = FALSE. Click **OK**. The function returns 0.1804. Now select the **BINOM.DIST** function. Click **OK**. Enter the following values: Number of successes = 3, Trials = 12, Probability of success = 0.1667, Cumulative = FALSE. Click **OK**. The function returns 0.1974. The Poisson approximation of 0.18 is too far from the correct result of 0.197.

Chapter Quick Quiz

1. Yes

2. $100 \cdot \dfrac{1}{5} = 20$

3. $\sigma = \sqrt{100 \cdot 0.2 \cdot 0.8} = 4$

4. The range of usual values has a minimum value of $200 - 2 \cdot 10 = 180$ and a maximum value of $200 + 2 \cdot 10 = 220$. Therefore, 232 girls in 400 is an unusually high number of girls since it is outside the range of usual values.

5. The range of usual values has a minimum value of $200 - 2 \cdot 10 = 180$ and a maximum value of $200 + 2 \cdot 10 = 220$. Therefore, 185 girls in 400 is not an unusually high number of girls since it is inside the range of usual values.

6. Yes. The sum of the probabilities is 0.999 and it can be considered to be 1.

7. 0+ indicates that the probability is a very small positive number. It does not indicate that it is impossible for none of the five flights to arrive on time.

8. $P(x \geq 3) = 0.198 + 0.409 + 0.338 = 0.945$

9. $\mu = 0 \cdot 0 + 1 \cdot 0.006 + 2 \cdot 0.048 + 3 \cdot 0.198 + 4 \cdot 0.409 + 5 \cdot 0.338 = 4.022$ and
    $\sigma = \sqrt{0^2 \cdot 0 + 1^2 \cdot 0.006 + 2^2 \cdot 0.048 + 3^2 \cdot 0.198 + 4^2 \cdot 0.409 + 5^2 \cdot 0.338 - 4.022^2} = 0.893$
    The range of usual values is from 2.236 to 5.808. Since zero is outside the range of usual values it is an unusually low number.

10. $\mu = 0 \cdot 0 + 1 \cdot 0.006 + 2 \cdot 0.048 + 3 \cdot 0.198 + 4 \cdot 0.409 + 5 \cdot 0.338 = 4.022$ and
    $\sigma = \sqrt{0^2 \cdot 0 + 1^2 \cdot 0.006 + 2^2 \cdot 0.048 + 3^2 \cdot 0.198 + 4^2 \cdot 0.409 + 5^2 \cdot 0.338 - 4.022^2} = 0.893$
    The range of usual values is from 2.236 to 5.808. Since 5 is inside the range of usual values it is not an unusually high number.

Review Exercises

1. You will use Excel's BINOM.DIST function to solve the problem. Click **Formulas** at the top of the screen. Select **Insert Function**. Select the **Statistical** category. Select the **BINOM.DIST** function. Click **OK**. Enter the following values: Number of successes = 0, Trials = 6, Probability of success = 0.4, Cumulative = FALSE. Click **OK**. The function returns 0.0467.

2. You will use Excel's BINOM.DIST function to solve the problem. Click **Formulas** at the top of the screen. Select **Insert Function**. Select the **Statistical** category. Select the **BINOM.DIST** function. Click **OK**. Enter the following values: Number of successes = 4, Trials = 6, Probability of success = 0.4, Cumulative =   FALSE. Click **OK**. The function returns 0.1382.

3. $\mu = 600 \cdot 0.4 = 240$ and $\sigma = \sqrt{600 \cdot 0.4 \cdot 0.6} = 12$. The range of usual values has a minimum of $240 - 2 \cdot 12 = 216$ and a maximum value of $240 + 2 \cdot 12 = 264$. The result of 200 with brown eyes is unusually low.

4. The probability of 239 or fewer (0.484) is relevant for determining whether 239 is an unusually low number. Because that probability is not very small, it appears that 239 is not an unusually low number of people with brown eyes.

5. Yes. The three requirements are satisfied. There is a numerical random variable x and its values are associated with corresponding probabilities. The sum of the probabilities is 1.001, so the sum is 1 when we allow for a small discrepancy due to rounding. Also, each of the probability values is between 0 and 1 inclusive.

6. $\mu = 0 \cdot 0.674 + 1 \cdot 0.28 + 2 \cdot 0.044 + 3 \cdot 0.003 + 4 \cdot 0 = 0.4$ and
    $\sigma = \sqrt{0^2 \cdot 0.674 + 1^2 \cdot 0.28 + 2^2 \cdot 0.044 + 3^2 \cdot 0.003 + 4^2 \cdot 0 - 0.4^2} = 0.6$
    The range of usual values has a minimum of $0.4 - 2 \cdot 0.6 = -0.8$ and a maximum value of $0.4 + 2 \cdot 0.6 = 1.6$. Yes, 3 is an unusually high number of males with tinnitus among four randomly selected males.

7. The sum of the probabilities is 0.902 which is not 1 as required. Because the three requirements are not satisfied, the given information does not describe a probability distribution.

8. Construct an Excel worksheet like the one shown below on the left with prize amounts in column B and probabilities in column C. Column D contains formulas to multiply column B amounts by column D amounts. The formulas are displayed in the worksheet at the left, and the results are displayed in the worksheet at the right. Mean = $\mu = \sum x \cdot P(x) = \$315,075$. Because the offer is well below her expected value, she should continue the game (although the guaranteed prize of $193000 had considerable appeal).

	A	B	C	D
1		x	P(x)	x*P(x)
2		75	=1/5	=B2*C2
3		300	=1/5	=B3*C3
4		75000	=1/5	=B4*C4
5		500000	=1/5	=B5*C5
6		1000000	=1/5	=B6*C6
7	Sum			=SUM(D2:D6)

	A	B	C	D
1		x	P(x)	x*P(x)
2		75	0.2	15
3		300	0.2	60
4		75000	0.2	15000
5		500000	0.2	100000
6		1000000	0.2	200000
7	Sum			315075

9. a. Construct an Excel worksheet like the one shown below on the left with prize amounts in column B and probabilities in column C. Column D contains formulas to multiply column B amounts by column D amounts. The formulas are displayed in the worksheet on the left, and the results are displayed in the worksheet on the right. Mean = $\mu = \sum x \cdot P(x) = \$0.012$.

	A	B	C	D
1		x	P(x)	x*P(x)
2		1000000	=1/90000000	=B2*C2
3		100000	=1/110000000	=B3*C3
4		25000	=1/110000000	=B4*C4
5		5000	=1/36667000	=B5*C5
6		2500	=1/27500000	=B6*C6
7	Sum			=SUM(D2:D6)

	A	B	C	D
1		x	P(x)	x*P(x)
2		1000000	1.11E-08	0.011111
3		100000	9.09E-09	0.000909
4		25000	9.09E-09	0.000227
5		5000	2.73E-08	0.000136
6		2500	3.64E-08	9.09E-05
7	Sum			0.012475

b. $0.012 minus the cost of the stamp. Since the expected value of winning is much smaller than the cost of a postage stamp, it is not worth entering the contest.

10. a. $\mu = \dfrac{18}{30} = 0.6$

b. You will use Excel's POISSON.DIST function to solve the problem. Click **Formulas** at the top of the screen. Select **Insert Function**. Select the **Statistical** category. Select the **POISSON.DIST** function. Click **OK**. Enter the following values: x = 0, Mean = 0.6, Cumulative = FALSE. Click **OK**. The function returns 0.5488.

c. $(30)(0.5488) = 16.464$

d. The expected number of days is 16.5, and that is reasonably close to the actual number of days which is 18.

Cumulative Review Exercises

1. Type the label **Hours** in an Excel worksheet followed by the data given in the problem. Select the **XLSTAT** add-in. Select **Describing data**. Select **Descriptive statistics**. Below Quantitative data, enter the data range inclusive of the label. Select **Sample labels**. Click the **Options** tab. Select **Descriptive Statistics**. Click the **Outputs** tab. Select **Range, Median, Mean, Variance (n-1)**, and **Standard deviation (n-1)**. Click **OK**.

Statistic	Hours
Range	4.7000
Median	24.2000
Mean	24.3800
Variance (n-1)	2.9120
Standard deviation (n-1)	1.7065

a. The mean is 24.38 hours.
b. The median is 24.2 hours.
c. The range is 4.7 hours.

    d.   The standard deviation is 1.7065 hours.

    e.   The variance is 2.9120 hours squared.

    f.   The minimum is $24.4 - 2 \cdot 1.7 = 21$ hours and the maximum is $24.4 + 2 \cdot 1.7 = 27.8$ hours.

    g.   No, because none of the times are outside the range of the usual values.

    h.   Ratio

    i.   Continuous

    j.   The given times come from countries with very different population sizes, so it does not make sense to treat the given times equally. Calculations of statistics should take the different population sizes into account. Also, the sample is very small, and there is no indication that the sample is random.

2.   a. $\dfrac{1}{10} \cdot \dfrac{1}{10} \cdot \dfrac{1}{10} \cdot \dfrac{1}{10} = \dfrac{1}{10,000} = 0.0001$

    b.

x	P(x)
−$1	0.9999
$4999	0.0001

    c. $365 \cdot 0.0001 = 0.0365$

    d.   You will use Excel's POISSON.DIST function to solve the problem. Click **Formulas** at the top of the screen. Select **Insert Function**. Select the **Statistical** category. Select the **POISSON.DIST** function. Click **OK**. Enter the following values: x = 1, Mean = .0365, Cumulative = FALSE. Click **OK**. The function returns 0.0352.

    e. $-\$1 \cdot 0.9999 + \$4999 \cdot 0.0001 = -\$0.50$ or -50 cents.

3.   a. $\dfrac{121 + 51}{611} = 0.282$

    b. $\dfrac{121}{121 + 279} = 0.303$

    c. $\dfrac{51}{51 + 160} = 0.242$

    d. $\dfrac{51}{51 + 121} = 0.297$

    e. $\dfrac{121 + 51}{611} \cdot \dfrac{121 + 51}{611} = 0.0792$

    f. $\dfrac{121 + 279}{611} + \dfrac{121 + 51}{611} - \dfrac{121}{611} = 0.738$

    g. $\dfrac{\left(\dfrac{121}{611}\right)}{\left(\dfrac{172}{611}\right)} = 0.703$

4.   Because the vertical scale begins at 60 instead of 0, the difference between the two amounts is exaggerated. The graph makes it appear that men's earnings are roughly twice those of women, but men earn roughly 1.2 times the earnings of women.

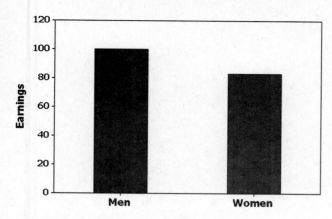

5. a.   Frequency distribution or frequency table

    b.   Probability distribution

    c. $\bar{x} = \dfrac{0\cdot9 + 1\cdot7 + 2\cdot12 + 3\cdot10 + 4\cdot10 + 5\cdot11 + 6\cdot8 + 7\cdot8 + 8\cdot14 + 9\cdot11}{9 + 7 + 12 + 10 + 10 + 11 + 8 + 8 + 14 + 11} = 4.7$

         This value is a statistic

    d. $\mu = 0\cdot0.1 + 1\cdot0.1 + 2\cdot0.1 + 3\cdot0.1 + 4\cdot0.1 + 5\cdot0.1 + 6\cdot0.1 + 7\cdot0.1 + 8\cdot0.1 + 9\cdot0.1 = 4.5$. This value is a
         parameter

    e.   The random generation of 1000 digits should have a mean close to 4.5 from part (d). The mean of 4.5 is the
         mean for the population of all random digits; so samples will have means that tend to center about 4.5

6. a. $_{16}C_4 \cdot 0.1^4 \cdot 0.9^{12} = 0.0514$

         You can use Excel's BINOM.DIST function to solve the problem. Click **Formulas** at the top of the screen.
         Select **Insert Function**. Select the **Statistical** category. Select the **BINOM.DIST** function. Click **OK**.
         Enter the following values: Number of successes = 4, Trials = 16, Probability of success = 0.1, Cumulative
         = FALSE. Click **OK**. The function returns 0.0514.

    b. $1 - {_{16}C_0} \cdot 0.1^0 \cdot 0.9^{16} = 0.815$

         You will find the probability that none believe that college is no longer a good investment, then you will
         subtract this result from 1. Select the **BINOM.DIST** function. Click **OK**. Enter the following values:
         Number of successes = 0, Trials = 16, Probability of success = 0.1, Cumulative = FALSE. Click **OK**. The
         function returns 0.1853. $1 - 0.1853 = 0.8147$

    c.   This is a voluntary response sample. This suggests that the results might not be valid, because those with a
         strong interest in the topic are more likely to respond.

Chapter 6

Normal Probability Distributions

Section 6-2, Basic Skills and Concepts

1. The word "normal" has a special meaning in statistics. It refers to a specific bell-shaped distribution that can be described by Formula 6-1.

3. The mean and standard deviation have values of $\mu = 0$ and $\sigma = 1$

5. $0.2(5 - 1.25) = 0.75$

7. $0.2(3 - 1) = 0.40$

9. You will use Excel's NORM.S.DIST function to solve the problem. Click **Formulas** at the top of the screen. Select **Insert Function**. Select the **Statistical** category. Select the **NORM.S.DIST** function. Click **OK**. Enter the following values: $z = 0.44$, Cumulative = TRUE. Click **OK**. The function returns 0.6700.
   $P(z < 0.44) = 0.6700$

11. You will use Excel's NORM.S.DIST function to solve the problem. You will first find the cumulative area to the left of $z = 1.28$. Next you will find the cumulative area to the left of $z = -0.84$. Then you will subtract the cumulative area to the left of $z = -0.84$ from the cumulative area to the left of $z = 1.28$. Click **Formulas** at the top of the screen. Select **Insert Function**. Select the **Statistical** category. Select the **NORM.S.DIST** function. Click **OK**. Enter the following values: $z = 1.28$, Cumulative = TRUE. Click **OK**. The function returns 0.8997. Select the **NORM.S.DIST** function again. Enter the following values: $z = -0.84$, Cumulative = TRUE. Click **OK**. The function returns 0.2005.
   $P(-0.84 < z < 1.28) = P(z < 1.28) - P(z < -0.84) = 0.8997 - 0.2005 = 0.6992$

13. You will use Excel's NORM.S.INV function to solve the problem. Click **Formulas** at the top of the screen. Select **Insert Function**. Select the **Statistical** category. Select the **NORM.S.INV** function. Enter the following value: Probability = 0.8907. The function returns 1.23. $z = 1.23$

15. You will use Excel's NORM.S.INV function to solve the problem. First subtract 0.9265 from 1. $1 - 0.9265 = 0.0735$. Click **Formulas** at the top of the screen. Select **Insert Function**. Select the **Statistical** category. Select the **NORM.S.INV** function. Enter the following value: Probability = 0.0735. The function returns −01.45.
   $z = -1.45$

17. You will use Excel's NORM.S.DIST function to solve the problem. Click **Formulas** at the top of the screen. Select **Insert Function**. Select the **Statistical** category. Select the **NORM.S.DIST** function. Enter the following values: $z = -2.04$, Cumulative = TRUE. The function returns 0.0207. $P(z < -2.04) = 0.0207$

19. You will use Excel's NORM.S.DIST function to solve the problem. Click **Formulas** at the top of the screen. Select **Insert Function**. Select the **Statistical** category. Select the **NORM.S.DIST** function. Click **OK**. Enter the following values: $z = 2.33$, Cumulative = TRUE. Click **OK**. The function returns 0.9901.
   $P(z < 2.33) = 0.9901$

21. You will use Excel's NORM.S.DIST function to solve the problem. Click **Formulas** at the top of the screen. Select **Insert Function**. Select the **Statistical** category. Select the **NORM.S.DIST** function. Enter the following values: $z = 0.82$, Cumulative = TRUE. The function returns 0.7939. $1 - 0.7939 = 0.2061$.
   $P(z > 0.82) = 1 - 0.7939 = 0.2061$

23. You will use Excel's NORM.S.DIST function to solve the problem. Click **Formulas** at the top of the screen. Select **Insert Function**. Select the **Statistical** category. Select the **NORM.S.DIST** function. Enter the following values: $z = -1.50$, Cumulative = TRUE. The function returns 0.0668. $1 - 0.0668 = 0.9332$.
   $P(z > -1.50) = 1 - 0.0668 = 0.9332$

25. You will use Excel's NORM.S.DIST function to solve the problem. You will first find the cumulative area to the left of $z = 1.25$. Next you will find the cumulative area to the left of $z = 0.25$. Then you will subtract the cumulative area to the left of $z = 0.25$ from the cumulative area to the left of $z = 1.25$. Click **Formulas** at the top of the screen. Select **Insert Function**. Select the **Statistical** category. Select the **NORM.S.DIST** function. Enter the following values: $z = 1.25$, Cumulative = TRUE. The function returns 0.8944.

    Select the **NORM.S.DIST** function again. Enter the following values: $z = 0.25$, Cumulative = TRUE. The function returns 0.5987. $P(0.25 < z < 1.25) = P(z < 1.25) - P(z < 0.25) = 0.8944 - 0.5987 = 0.2957$

27. You will use Excel's NORM.S.DIST function to solve the problem. You will first find the cumulative area to the left of $z = -2.00$. Next you will find the cumulative area to the left of $z = -2.75$. Then you will subtract the cumulative area to the left of $z = -2.00$ from the cumulative area to the left of $z = -2.75$. Click **Formulas** at the top of the screen. Select **Insert Function**. Select the **Statistical** category. Select the **NORM.S.DIST** function. Enter the following values: $z = -2.00$, Cumulative = TRUE. The function returns 0.0228.

    Select the **NORM.S.DIST** function again. Enter the following values: $z = -2.75$, Cumulative = TRUE. The function returns 0.0030. $P(-2.75 < z < -2.00) = P(z < -2.00) - P(z < -2.75) = 0.0228 - 0.0030 = 0.0198$

29. You will use Excel's NORM.S.DIST function to solve the problem. You will first find the cumulative area to the left of $z = 2.50$. Next you will find the cumulative area to the left of $z = -2.20$. Then you will subtract the cumulative area to the left of $z = -2.20$ from the cumulative area to the left of $z = 2.50$. Click **Formulas** at the top of the screen. Select **Insert Function**. Select the **Statistical** category. Select the **NORM.S.DIST** function. Enter the following values: $z = 2.50$, Cumulative = TRUE. The function returns 0.9938.

    Select the **NORM.S.DIST** function again. Enter the following values: $z = -2.20$, Cumulative = TRUE. The function returns 0.0139. $P(-2.20 < z < 2.50) = P(z < 2.50) - P(z < -2.20) = 0.9938 - 0.0139 = 0.9799$

31. You will use Excel's NORM.S.DIST function to solve the problem. You will first find the cumulative area to the left of $z = 4.00$. Next you will find the cumulative area to the left of $z = -2.11$. Then you will subtract the cumulative area to the left of $z = -2.11$ from the cumulative area to the left of $z = 4.00$. Click **Formulas** at the top of the screen. Select **Insert Function**. Select the **Statistical** category. Select the **NORM.S.DIST** function. Enter the following values: $z = 4.00$, Cumulative = TRUE. The function returns 0.999968.

    Select the **NORM.S.DIST** function again. Enter the following values: $z = -2.11$, Cumulative = TRUE. The function returns 0.017429.
    $$P(-2.11 < z < 4.00) = P(z < 4.00) - P(z < -2.11) = 0.999968 - 0.017429 = 0.9825$$

33. You will use Excel's NORM.S.DIST function to solve the problem. Click **Formulas** at the top of the screen. Select **Insert Function**. Select the **Statistical** category. Select the **NORM.S.DIST** function. Enter the following values: $z = 3.65$, Cumulative = TRUE. The function returns 0.999869. $P(z < 3.65) = 0.9999$

35. You will use Excel's NORM.S.DIST function to solve the problem. Click **Formulas** at the top of the screen. Select **Insert Function**. Select the **Statistical** category. Select the **NORM.S.DIST** function. Enter the following values: $z = 0$, Cumulative = TRUE. The function returns 0.5000. $P(z < 0) = 0.5000$

37. You will use Excel's NORM.S.INV function to solve the problem. Click **Formulas** at the top of the screen. Select **Insert Function**. Select the **Statistical** category. Select the **NORM.S.INV** function. Enter the following value: Probability = 0.90. The function returns 1.28. $P_{90} = 1.28$

39. You will use Excel's NORM.S.INV function to solve the problem. Click **Formulas** at the top of the screen. Select **Insert Function**. Select the **Statistical** category. Select the **NORM.S.INV** function. Enter the following value: Probability = 0.025. The function returns -1.96. Because the distribution is symmetrical, $P_{2.5} = -1.96$ and $P_{97.5} = 1.96$

41. You will use Excel's NORM.S.INV function to solve the problem. Click **Formulas** at the top of the screen. Select **Insert Function**. Select the **Statistical** category. Select the **NORM.S.INV** function. Enter the following value: Probability = 0.025. The function returns -1.96. $z_{0.025} = 1.96$

43. You will use Excel's NORM.S.INV function to solve the problem. Click **Formulas** at the top of the screen. Select **Insert Function**. Select the **Statistical** category. Select the **NORM.S.INV** function. Enter the following value: Probability = 0.05. The function returns −1.645. $z_{0.05} = 1.645$

45. You will use Excel's NORM.S.DIST function to solve the problem. You will first find the cumulative area to the left of $z = 1$. Next you will find the cumulative area to the left of $z = -1$. Then you will subtract the cumulative area to the left of $z = -1$ from the cumulative area to the left of $z = 1$. Click **Formulas** at the top of the screen. Select **Insert Function**. Select the **Statistical** category. Select the **NORM.S.DIST** function. Enter the following values: $z = 1$, Cumulative = TRUE. The function returns 0.8413.

   Select the **NORM.S.DIST** function again. Enter the following values: $z = -1$, Cumulative = TRUE. The function returns 0.1587. $P(-1 < z < 1) = P(z < 1) - P(z < -1) = 0.8413 - 0.1587 = 0.6826 = 68.26\%$

47. You will use Excel's NORM.S.DIST function to solve the problem. You will first find the cumulative area to the left of $z = 3$. Next you will find the cumulative area to the left of $z = -3$. Then you will subtract the cumulative area to the left of $z = -3$ from the cumulative area to the left of $z = 3$. Click **Formulas** at the top of the screen. Select **Insert Function**. Select the **Statistical** category. Select the **NORM.S.DIST** function. Enter the following values: $z = 3$, Cumulative = TRUE. The function returns 0.9987.

   Select the **NORM.S.DIST** function again. Enter the following values: $z = -3$, Cumulative = TRUE. The function returns 0.0013. $P(-3 < z < 3) = P(z < 3) - P(z < -3) = 0.9987 - 0.0013 = 0.9974 = 99.74\%$

Section 6-2, Beyond the Basics

49. a. You will use Excel's NORM.S.DIST function to solve the problem. You will first find the cumulative area to the left of $z = 1$. Next you will find the cumulative area to the left of $z = -1$. Then you will subtract the cumulative area to the left of $z = -1$ from the cumulative area to the left of $z = 1$. Click **Formulas** at the top of the screen. Select **Insert Function**. Select the **Statistical** category. Select the **NORM.S.DIST** function. Enter the following values: $z = 1$, Cumulative = TRUE. The function returns 0.8413.

   Select the **NORM.S.DIST** function again. Enter the following values: $z = -1$, Cumulative = TRUE. The function returns 0.1587. $P(-1 < z < 1) = P(z < 1) - P(z < -1) = 0.8413 - 0.1587 = 0.6826 = 68.26\%$

   b. You will use Excel's NORM.S.DIST function to solve the problem. You will find the cumulative area to the left of $z = -2$. Click **Formulas** at the top of the screen. Select **Insert Function**. Select the **Statistical** category. Select the **NORM.S.DIST** function. Enter the following values: $z = -2$, Cumulative = TRUE. The function returns 0.0228. Because the distribution is symmetrical, $P(z < -2 \text{ or } z > 2) = P(z < -2) + P(z > 2) = 0.0228 + 0.0228 = 0.0456 = 4.56\%$

   c. You will use Excel's NORM.S.DIST function to solve the problem. You will find the cumulative area to the left of $z = -1.96$. Click **Formulas** at the top of the screen. Select **Insert Function**. Select the **Statistical** category. Select the **NORM.S.DIST** function. Enter the following values: $z = -1.96$, Cumulative = TRUE. The function returns 0.0250. Because the distribution is symmetrical, $P(z < -1.96 \text{ or } z > 1.96) = P(z < -1.96) + P(z > 1.96) = 0.0250 + 0.0250 = 0.0500$
   $P(-1.96 < z < 1.96) = 1 - 0.0500 = 0.9500 = 95\%$

   d. You will use Excel's NORM.S.DIST function to solve the problem. You will first find the cumulative area to the left of $z = 2$. Next you will find the cumulative area to the left of $z = -2$. Then you will subtract the cumulative area to the left of $z = -2$ from the cumulative area to the left of $z = 2$. Click **Formulas** at the top of the screen. Select **Insert Function**. Select the **Statistical** category.

   Select the **NORM.S.DIST** function. Enter the following values: $z = 2$, Cumulative = TRUE. The function returns 0.9772. Select the **NORM.S.DIST** function again. Enter the following values: $z = -2$, Cumulative = TRUE. The function returns 0.0228.
   $P(-2 < z < 2) = P(z < 2) - P(z < -2) = 0.9772 - 0.0228 = 0.9544 = 95.44\%$

e.   You will use Excel's NORM.S.DIST function to solve the problem. You will first find the cumulative area to the left of $z = -3$. Click **Formulas** at the top of the screen. Select **Insert Function**. Select the **Statistical** category. Select the **NORM.S.DIST** function. Enter the following values: $z = -3$, Cumulative = TRUE. The function returns 0.0013. Because the distribution is symmetrical,

$$P(z < -3 \text{ or } z > 3) = P(z < -3) + P(z > 3) = 0.0013 + 0.0013 = 0.0026 = 0.26\%$$

Section 6-3, Basic Skills and Concepts

1.   a. $\mu = 0$ and $\sigma = 1$.

     b.   The z scores are numbers without units of measurements.

3.   The standard normal distribution has a mean of 0 and a standard deviation of 1, but a nonstandard normal distribution has a different value for one or both of those parameters.

5.   You will use Excel's STANDARDIZE function to find z. Click **FORMULAS** at the top of the screen. Select **Insert Function**. Select the **Statistical** category. Select the **STANDARDIZE** function. Enter the following values: x = 118, Mean = 100, Standard deviation = 15. Click **OK**. The function returns 1.2; $z_{x=118} = 1.2$. Next, you will use the NORM.S.DIST function to find the area. Select the **Statistical** category. Select the **NORM.S.DIST** function. Enter the following values: $z = 1.2$, Cumulative = TRUE. The function returns 0.8848. The area to the left of $z = 1.2$ is 0.8848.

7.   You will use Excel's STANDARDIZE function to find the two z values. Click **FORMULAS** at the top of the screen. Select **Insert Function**. Select the **Statistical** category. Select the **STANDARDIZE** function. Enter the following values: x = 79, Mean = 100, Standard deviation = 15. The function returns –1.4; $z_{x=79} = -1.4$.

Select the **STANDARDIZE** function again. Enter the following values: x = 133, Mean = 100, Standard deviation = 15. The function returns –2.2; $z_{x=133} = 2.2$. Next, you will use the NORM.S.DIST function to find the area. Select the **Statistical** category. Select the **NORM.S.DIST** function. Enter the following values: $z = -1.4$, Cumulative = TRUE. The function returns 0.0808. The area to the left of $z = -1.4$ is 0.0808.

Select the **NORM.S.DIST** function again. Enter the following values: $z = 2.2$, Cumulative = TRUE. The function returns 0.9861. The area between the two scores is $0.9861 - 0.0808 = 0.9053$.

9.   You will use Excel's NORM.INV function to find the IQ score. Click **FORMULAS** at the top of the screen. Select **Insert Function**. Select the **Statistical** category. Select the **NORM.INV** function. Enter the following values: Probability = 0.9918, Mean = 100, Standard deviation = 15. The function returns an IQ score of 136.

11.   First find the area to the left of x: $1 - 0.9798 = 0.0202$. Now you will use Excel's NORM.INV function to find the IQ score. Click **FORMULAS** at the top of the screen. Select **Insert Function**. Select the **Statistical** category. Select the **NORM.INV** function. Enter the following values: Probability = 0.0202, Mean = 100, Standard deviation = 15. The function returns an IQ score of 69.

13.   You will use Excel's NORM.DIST function to find the probability. Click **FORMULAS** at the top of the screen. Select **Insert Function**. Select the **Statistical** category. Select the **NORM.DIST** function. Enter the following values: x = 85, Mean = 100, Standard deviation = 15, Cumulative = TRUE. The function returns a probability of 0.1587.

15.   You will use Excel's NORM.DIST function to find the probability. Click **FORMULAS** at the top of the screen. Select **Insert Function**. Select the **Statistical** category. Select the **NORM.DIST** function. Enter the following values: x = 90, Mean = 100, Standard deviation = 15, Cumulative = TRUE. The function returns a probability of 0.2525.

Select the **NORM.DIST** function again. Enter the following values: x = 110, Mean = 100, Standard deviation = 15, Cumulative = TRUE. The function returns a probability of 0.7475. The area between the two IQ scores $= 0.7475 - 0.2525 = 0.4950$.

17. You will use Excel's NORM.INV function to find the IQ score. Click **FORMULAS** at the top of the screen. Select **Insert Function**. Select the **Statistical** category. Select the **NORM.INV** function. Enter the following values: Probability = 0.9, Mean = 100, Standard deviation = 15. The function returns an IQ score of 119.

19. You will use Excel's NORM.INV function to find the IQ score. Click **FORMULAS** at the top of the screen. Select **Insert Function**. Select the **Statistical** category. Select the **NORM.INV** function. Enter the following values: Probability = 0.75, Mean = 100, Standard deviation = 15. The function returns an IQ score of 110.

21. a.   You will use Excel's NORM.DIST function to solve the problem. Click **FORMULAS** at the top of the screen. Select **Insert Function**. Select the **Statistical** category. Select the **NORM.DIST** function. Enter the following values: x = 62, Mean = 63.8, Standard deviation = 2.6, Cumulative = TRUE. The function returns a probability of 0.2444.

   Select the **NORM.DIST** function again. Enter the following values: x = 78, Mean = 63.8, Standard deviation = 2.6, Cumulative = TRUE. The function returns a probability of 0.999999976. The area between 62 and 78 in. =0.999999976−0.2444=0.7556 or 75.56%.

   b.   You will use Excel's NORM.DIST function to solve the problem. Click **FORMULAS** at the top of the screen. Select **Insert Function**. Select the **Statistical** category. Select the **NORM.DIST** function. Enter the following values: x = 62, Mean = 69.5, Standard deviation = 2.4, Cumulative = TRUE. The function returns a probability of 0.0009.

   Select the **NORM.DIST** function again. Enter the following values: x = 78, Mean = 69.5, Standard deviation = 2.4, Cumulative = TRUE. The function returns a probability of 0.9998. The area between 62 and 78 in. =0.9998−0.0009=0.9989 or 99.89%. No, only about 0.1% of men are not qualified because of their heights.

   c.   You will use Excel's NORM.INV function to find the heights. Click **FORMULAS** at the top of the screen. Select **Insert Function**. Select the **Statistical** category. Select the **NORM.INV** function. Enter the following values: Probability = 0.02, Mean = 63.8, Standard deviation = 2.6. The function returns height of 58.5 inches.

   Select the **NORM.INV** function again. Enter the following values: Probability = 0.98, Mean = 63.8, Standard deviation = 2.6. The function returns height of 69.1 inches. The new height requirements are 58.5 inches to 69.1 inches.

   d.   You will use Excel's NORM.INV function to find the heights. Click **FORMULAS** at the top of the screen. Select **Insert Function**. Select the **Statistical** category. Select the **NORM.INV** function. Enter the following values: Probability = 0.01, Mean = 69.5, Standard deviation = 2.4. The function returns height of 63.9 inches.

   Select the **NORM.INV** function again. Enter the following values: Probability = 0.99, Mean = 69.5, Standard deviation = 2.4. The function returns height of 75.1 inches. The new height requirements are 63.9 inches to 75.1 inches.

23. a.   You will use Excel's NORM.DIST function to solve the problem. Click **FORMULAS** at the top of the screen. Select **Insert Function**. Select the **Statistical** category. Select the **NORM.DIST** function. Enter the following values: x = 56, Mean = 63.8, Standard deviation = 2.6, Cumulative = TRUE. The function returns a probability of 0.0013.

   Select the **NORM.DIST** function again. Enter the following values: x = 75, Mean = 63.8, Standard deviation = 2.6, Cumulative = TRUE. The function returns a probability of 0.9999918. The area between 56 and 75 in. =0.9999918−0.0013=0.9987 or 99.87%.

   b.   You will use Excel's NORM.DIST function to solve the problem. Click **FORMULAS** at the top of the screen. Select **Insert Function**. Select the **Statistical** category. Select the **NORM.DIST** function. Enter the following values: x = 56, Mean = 69.5, Standard deviation = 2.4, Cumulative = TRUE. The function returns a probability of 0.000000009.

Select the **NORM.DIST** function again. Enter the following values: x = 75, Mean = 69.5, Standard deviation = 2.4, Cumulative = TRUE. The function returns a probability of 0.9890. The area between 56 and 75 in. $=0.9890-0.000000009=0.9890$ or 98.90%.

c.   You will use Excel's NORM.INV function to find the heights. Click **FORMULAS** at the top of the screen. Select **Insert Function**. Select the **Statistical** category. Select the **NORM.INV** function. Enter the following values: Probability = 0.05, Mean = 63.8, Standard deviation = 2.6. The function returns height of 59.5 inches.

Select the **NORM.INV** function again. Enter the following values: Probability = 0.97, Mean = 69.5, Standard deviation = 2.4. The function returns height of 73.4 inches. The new height requirements are 59.5 inches to 73.4 inches.

25. a.   You will use Excel's NORM.DIST function to solve the problem. Click **FORMULAS** at the top of the screen. Select **Insert Function**. Select the **Statistical** category. Select the **NORM.DIST** function. Enter the following values: x = 174, Mean = 182.9, Standard deviation = 40.8, Cumulative = TRUE. The function returns a probability of 0.4137.

b. $\dfrac{3500}{140} = 25$ people

c. $\dfrac{3500}{182.9} = 19.14$, so 19 people

d.   The mean weight is increasing over time, so safety limits must be periodically updated to avoid an unsafe condition.

27. a.   You will use Excel's NORM.DIST function to solve the problem. Click **FORMULAS** at the top of the screen. Select **Insert Function**. Select the **Statistical** category. Select the **NORM.DIST** function. Enter the following values: x = 308 Mean = 268, Standard deviation = 15, Cumulative = TRUE. The function returns a probability of 0.9962. The probability of a pregnancy lasting 308 days or longer is $1-0.9962=0.0038$. Either a very rare event occurred or the husband is not the father.

b.   You will use Excel's NORM.INV function to solve the problem. Click **FORMULAS** at the top of the screen. Select **Insert Function**. Select the **Statistical** category. Select the **NORM.INV** function. Enter the following values: Probability = .03, Mean = 268, Standard deviation = 15. The function returns 239.8. The length that separates premature from not premature is 240 days.

29. a.   You will use Excel's NORM.DIST function to solve the problem. Click **FORMULAS** at the top of the screen. Select **Insert Function**. Select the **Statistical** category. Select the **NORM.DIST** function. Enter the following values: x = 2 Mean = 1.184, Standard deviation = 0.587, Cumulative = TRUE. The function returns a probability of 0.9178. 91.78% of earthquakes are considered microearthquakes.

b.   You will use Excel's NORM.DIST function to solve the problem. Click **FORMULAS** at the top of the screen. Select **Insert Function**. Select the **Statistical** category. Select the **NORM.DIST** function. Enter the following values: x = 4 Mean = 1.184, Standard deviation = 0.587, Cumulative = TRUE. The function returns a probability of 0.999999. $1-0.999999=0.000001=0.00\%$ of earthquakes fall into this category.

c.   You will use Excel's NORM.INV function to solve the problem. Click **FORMULAS** at the top of the screen. Select **Insert Function**. Select the **Statistical** category. Select the **NORM.INV** function. Enter the following values: Probability = 0.95, Mean = 1.184, Standard deviation = 0.587. The function returns 2.15. Not all earthquakes about the 95th percentile will cause items to shake.

31.   You will use Excel's NORM.INV function to solve the problem. Click **FORMULAS** at the top of the screen. Select **Insert Function**. Select the **Statistical** category. Select the **NORM.INV** function. Enter the following values: Probability = 0.01, Mean = 24, Standard deviation = 2.6. The function returns 18.0; $P_1 = 18$ chocolate chips.

Select the **NORM.INV** function again. Enter the following values: Probability = 0.99, Mean = 24, Standard deviation = 2.6. The function returns 30.0; $P_{99} = 30$ chocolate chips. The values can be used to identify cookies with an unusually low number of chocolate chips or an unusually high number of chocolate chips, so those numbers can be used to monitor the production process to ensure that the numbers of chocolate chips stay within reasonable limits.

33. a.   Open the **MBODY** data file on your data disk. Select the **XLSTAT** add-in. Select **Visualizing data**. Select **Histograms**. Enter the pulse data range inclusive of the label. Select **Discrete**. Select **Sample labels**. Click the **Outputs** tab. Select **Descriptive statistics**. Click the **Charts** tab. Select **Histograms** and **Bars**. Click **OK**. Mean = 67.25 beats per minute. Standard deviation = 10.335 beats per minute. The histogram, shown below, confirms that the distribution is roughly normal.

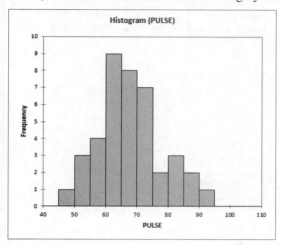

b.   You will use Excel's NORM.INV function to solve the problem. Click **FORMULAS** at the top of the screen. Select **Insert Function**. Select the **Statistical** category. Select the **NORM.INV** function. Enter the following values: Probability = 0.025, Mean = 67.25, Standard deviation = 10.335. The function returns 46.99; $P_{2.5}$ = 47.0 beats per minute g.

Select the **NORM.INV** function again. Enter the following values: Probability = 0.975, Mean = 67.25, Standard deviation = 10.335. Click **OK**. The function returns 87.51; $P_{97.5}$ = 87.5 beats per minute.

Section 6-3, Beyond the Basics

35. a.   The new mean is equal to the old one plus the new points which is 75. The standard deviation is unchanged at 10 (since we added the same amount to each student.)

b.   No, the conversion should also account for variation.

c.   You will use Excel's NORM.INV function to solve the problem. Click **FORMULAS** at the top of the screen. Select **Insert Function**. Select the **Statistical** category. Select the **NORM.INV** function. Enter the following values: Probability = 0.70, Mean = 40, Standard deviation = 10. The function returns 45.2.

Select the **NORM.INV** function again. Enter the following values: Probability = 0.90, Mean = 40, Standard deviation = 10. The function returns 52.8. B grade: 45.2 to 52.8.

d.   Using a scheme like the one in part (c), because variation is included in the curving process.

37.   The z score for Q_1 is –0.67, and the z score for Q_3 is 0.67. The IQR is $0.67 - (-0.67) = 1.34$. $1.5 \cdot IQR = 2.01$ so $Q_1 - 1.5 \cdot IQR = -0.67 - 2.01 = -2.68$ and $Q_3 + 1.5 \cdot IQR = 0.67 + 2.01 = 2.68$.

The percentage to the left of –2.68 is 0.0037 and the percentage to the right of 2.68 is 0.0037. Therefore, the probability of an outlier is 0.0074.

Section 6-4, Basic Skills and Concepts

1. a.   The sample mean will tend to center about the population parameter of 5.67 g.

    b.   The sample mean will tend to have a distribution that is approximately normal.

    c.   The sample proportions will tend to have a distribution that is approximately normal.

3. Sample mean, sample variance, sample proportion

5. No. The sample is not a simple random sample from the population of all college Statistics students. It is very possible that the students at Broward College do not accurately reflect the behavior of all college Statistics students.

7. a.   The mean of the population is $\mu = \dfrac{4+5+9}{3} = 6$, and the variance is

    $$\sigma^2 = \dfrac{(4-6)^2 + (5-6)^2 + (9-6)^2}{3} = 4.7$$

    b.   The possible sample of size 2 are {(4, 4), (4, 5), (4, 9), (5, 4), (5, 5), (5, 9), (9, 4), (9, 5), (9, 9)} which have the following variances {0, 0.5, 12.5, 0.5, 0, 8, 12.5, 8, 0} respectively.

Sample Variance	Probability
0	3/9
0.5	2/9
8	2/9

    c.   The sample variances' mean is $\dfrac{3 \cdot 0 + 2 \cdot 0.5 + 2 \cdot 8 + 2 \cdot 12.5}{9} = 4.7$

    d.   Yes. The mean of the sampling distribution of the sample variances (4.7) is equal to the value of the population variance (4.7) so the sample variances target the value of the population variance.

9. a.   The population median is 5.

    b.   The possible sample of size 2 are {(4, 4), (4, 5), (4, 9), (5, 4), (5, 5), (5, 9), (9, 4), (9, 5), (9, 9)} which have the following medians {4, 4.5, 6.5, 4.5, 5, 7, 6.5, 7, 9}

Sample Median	Probability
4	1/9
4.5	2/9
5	1/9
6.5	2/9
7	2/9
9	1/9

    c.   The mean of the sampling distribution of the sampling median is
    $$\dfrac{4+4.5+4.5+5+6.5+6.5+7+7+9}{9} = 6$$

    d.   No. The mean of the sampling distribution of the sample medians is 6, and it is not equal to the value of the population median of 5, so the sample medians do not target the value of the population median.

11. a.   The possible samples of size 2 are {(56, 56), (56, 49), (56, 58), (56, 46), (49, 56), (49, 49), (49, 58), (49, 46), (58, 56), (58, 49), (58, 58), (58, 46), (46, 56), (46, 49), (46, 58), (46, 46)}

Sample Mean Age	Probability
46	1/16
47.5	2/16
49	1/16
51	2/16
52	2/16
52.5	2/16
53.5	2/16
56	1/16
57	2/16
58	1/16

b. The mean of the population is $\dfrac{56+49+58+46}{4}=52.25$ and the mean of the sample means is

$$\frac{46+47.5+47.5+49+51+51+52+52+52.5+52.5+53.5+53.5+56+57+57+58}{16}=52.25$$

c. The sample means target the population mean. Sample means make good estimators of population means because they target the value of the population mean instead of systematically underestimating or overestimating it.

13. a. The possible samples of size 2 are {(56, 56), (56, 49), (56, 58), (56, 46), (49, 56), (49, 49), (49, 58), (49, 46), (58, 56), (58, 49), (58, 58), (58, 46), (46, 56), (46, 49), (46, 58), (46, 46)} which have the following ranges and associated probabilities:

Sample Range	Probability
0	4/16
2	2/16
3	2/16
7	2/16
9	2/16
10	2/16
12	2/16

b. The range of the population is $58-46=12$, the mean of the sample ranges is
$\dfrac{4\cdot 0+2\cdot 2+2\cdot 3+2\cdot 7+2\cdot 9+2\cdot 10+2\cdot 12}{16}=5.375$. The values are not equal.

c. The sample ranges do not target the population range of 12, so sample ranges do not make good estimators of the population range.

15. The possible birth samples are $\{(b, b), (b, g), (g, b), (g, g)\}$.

Proportion of Girls	Probability
0	0.25
1 / 2	0.5
2/2	0.25

Yes. The proportion of girls in 2 births is 0.5, and the mean of the sample proportions is 0.5. The result suggests that a sample proportion is an unbiased estimator of the population proportion.

17. The possibilities are: both questions incorrect, one question correct (two choices), both questions correct.

a.

Proportion Correct	Probability
$\dfrac{0}{2}$	$\dfrac{4}{5} \cdot \dfrac{4}{5} = \dfrac{16}{25}$
$\dfrac{1}{2}$	$2 \cdot \left(\dfrac{1}{5} \cdot \dfrac{4}{5} \right) = \dfrac{8}{25}$
$\dfrac{2}{2}$	$\left(\dfrac{1}{5} \cdot \dfrac{1}{5} \right) = \dfrac{1}{25}$

b.   The mean is $\dfrac{16 \cdot 0 + 8 \cdot 0.5 + 1 \cdot 1}{25} = 0.2$.

c.   Yes. The sampling distribution of the sample proportions has a mean of 0.2 and the population proportion is also 0.2 (because there is 1 correct answer among 5 choices.) Yes, the mean of the sampling distribution of the sample proportions is always equal to the population proportion.

Section 6-4, Beyond the Basics

19. $P(0) = \dfrac{1}{2(2 - 2 \cdot 0)!(2 \cdot 0)!} = \dfrac{1}{4} = 0.25$, $P(0.5) = \dfrac{1}{2(2 - 2 \cdot 0.5)!(2 \cdot 0.5)!} = 0.5$, $P(1) = \dfrac{1}{2(2 - 2 \cdot 1)!(2 \cdot 1)!} = 0.25$.

The formula yields values which describes the sampling distribution of the sample proportions. The formula is just a different way of presenting the same information in the table that describes the sampling distribution.

Section 6-5, Basic Skills and Concepts

1.   Because the sample size is greater than 30, the sampling distribution of the mean ages can be approximated by a normal distribution with mean μ and standard deviation $\dfrac{\sigma}{\sqrt{40}}$.

3. $\mu_{\bar{x}} = 60.5$ cm and it represents the mean of the population consisting of all sample means. $\sigma_{\bar{x}} = \dfrac{6.6}{\sqrt{36}} = 1.1$ cm, and it represents the standard deviation of the population consisting of all sample means.

5.   a.   You will use Excel's NORM.DIST function to solve the problem. Click **FORMULAS** at the top of the screen. Select **Insert Function**. Select the **Statistical** category. Select the **NORM.DIST** function. Enter the following values: x = 222.7 Mean = 205.5, Standard deviation = 8.6, Cumulative = TRUE. The function returns a probability of 0.9772.

b. $\mu_{\bar{x}} = \mu = 205.5$, $\sigma_{\bar{x}} = \sigma / \sqrt{n} = 8.6 / \sqrt{49} = 1.2286$

Select the **NORM.DIST** function. Enter the following values: x = 207.0, Mean = 205.5, Standard deviation = 1.2286, Cumulative = TRUE. The function returns a probability of 0.8889.

7. a. You will use Excel's NORM.DIST function to solve the problem. Click **FORMULAS** at the top of the screen. Select **Insert Function**. Select the **Statistical** category. Select the **NORM.DIST** function. Enter the following values: x = 218.4 Mean = 205.5, Standard deviation = 8.6, Cumulative = TRUE. The function returns a probability of 0.9332. 1−0.9332=0.0668

b. $\mu_{\bar{x}} = \mu = 205.5$, $\sigma_{\bar{x}} = \sigma / \sqrt{n} = 8.6 / \sqrt{9} = 2.8667$

Select the **NORM.DIST** function. Enter the following values: x = 204, Mean = 205.5, Standard deviation = 0.9556, Cumulative = TRUE. The function returns a probability of 0.3004. 1−0.3004=0.6996

c. Because the original population has a normal distribution, the distribution of sample means is normal for any sample size.

9. a. You will use Excel's NORM.DIST function to solve the problem. Click **FORMULAS** at the top of the screen. Select **Insert Function**. Select the **Statistical** category. Select the **NORM.DIST** function. Enter the following values: x = 179.7 Mean = 205.5, Standard deviation = 8.6, Cumulative = TRUE. The function returns a probability of 0.0013.

Select the **NORM.DIST** function again. Enter the following values: x = 231.3, Mean = 205.5, Standard deviation = 8.6, Cumulative = TRUE. The function returns a probability of 0.9987. 0.9987−0.0013=0.9974

b. $\mu_{\bar{x}} = \mu = 205.5$, $\sigma_{\bar{x}} = \sigma / \sqrt{n} = 8.6 / \sqrt{40} = 1.3598$

Select the **NORM.DIST** function. Enter the following values: x = 204, Mean = 205.5, Standard deviation = 1.3598, Cumulative = TRUE. The function returns a probability of 0.1350.

Select the **NORM.DIST** function again. Enter the following values: x = 206 Mean = 205.5, Standard deviation = 1.3598, Cumulative = TRUE. The function returns a probability of 0.6435. 0.6435−0.1350=0.5085

11. $\mu_{\bar{x}} = \mu = 182.9$, $\sigma_{\bar{x}} = \sigma / \sqrt{n} = 40.8 / \sqrt{16} = 10.2$

Select the **NORM.DIST** function. Enter the following values: x = 195.3, Mean = 182.9, Standard deviation = 10.2, Cumulative = TRUE. The function returns a probability of 0.8879; 1−0.8879=0.1121. The elevator does not appear to be safe because there is a reasonable chance (0.1121) that it will be overloaded with 16 male passengers.

13. a. You will use Excel's NORM.DIST function to solve the problem. Click **FORMULAS** at the top of the screen. Select **Insert Function**. Select the **Statistical** category. Select the **NORM.DIST** function. Enter the following values: x = 21 Mean = 22.65, Standard deviation = 0.8, Cumulative = TRUE. The function returns a probability of 0.0196.

Select the **NORM.DIST** function again. Enter the following values: x = 25 Mean = 22.65, Standard deviation = 0.8, Cumulative = TRUE. The function returns a probability of 0.9983. 0.9983−0.0196=0.9787

b. You will use Excel's NORM.INV function to solve the problem. Click **FORMULAS** at the top of the screen. Select **Insert Function**. Select the **Statistical** category. Select the **NORM.INV** function. Enter the following values: Probability = 0.025, Mean = 22.65, Standard deviation = 0.8. The function returns a head size of 21.08 inches.

Select the **NORM.INV** function again. Enter the following values: Probability = 0.975, Mean = 22.65, Standard deviation = 0.8. The function returns a head size of 24.22 inches. 21.08 to 24.22 inches.

c. $\mu_{\bar{x}} = \mu = 22.65$, $\sigma_{\bar{x}} = \sigma / \sqrt{n} = 0.8 / \sqrt{64} = 0.1$

Select the **NORM.DIST** function. Enter the following values: x = 22, Mean = 22.65, Standard deviation = 0.1, Cumulative = TRUE. The function returns a probability of 4.016E−11 or 0.00000.

Select the **NORM.DIST** function again. Enter the following values: x = 23, Mean = 22.65, Standard deviation = 0.1, Cumulative = TRUE. The function returns a probability of 0.999767; $0.999767 - .00000 = 0.9998$. No, the hats must fit individual women, not the mean from 64 women. If all hats are made to fit head circumferences between 22.00 in. and 23.00 in., the hats won't fit about half of those women.

15. a.   The mean weight of passengers is $\dfrac{3500}{25} = 140$ lb.

    b. $\mu_{\bar{x}} = \mu = 182.9$, $\sigma_{\bar{x}} = \sigma / \sqrt{n} = 40.8 / \sqrt{25} = 8.16$

         Select the **NORM.DIST** function. Enter the following values: x = 140, Mean = 182.9, Standard deviation = 8.16, Cumulative = TRUE. The function returns a probability of 7.30718E–08 or 0.000000073. $1 - 0.000000073 = 1.0000$ rounded to 4 decimal places.

    c. $\sigma_{\bar{x}} = \sigma / \sqrt{n} = 40.8 / \sqrt{20} = 9.1232$

         Select the **NORM.DIST** function. Enter the following values: x = 175, Mean = 182.9, Standard deviation = 9.1232, Cumulative = TRUE. The function returns a probability of 0.1933. $1 - 0.1933 = 0.8067$

    d.   Given that there is a 0.8067 probability of exceeding the 3500 lb. limit when the water taxi is loaded with 20 random men, the new capacity of 20 passengers does not appear to be safe enough because the probability of overloading is too high.

17. a.   Select the **NORM.DIST** function. Enter the following values: x = 167, Mean = 182.9, Standard deviation 40.8, Cumulative = TRUE. The function returns a probability of 0.3484. $1 - 0.3484 = 0.6516$

    b. $\sigma_{\bar{x}} = \sigma / \sqrt{n} = 40.8 / \sqrt{12} = 11.7779$

         Select the **NORM.DIST** function. Enter the following values: x = 167, Mean = 182.9, Standard deviation 11.7779, Cumulative = TRUE. The function returns a probability of 0.0885. $1 - 0.0885 = 0.9115$

    c.   There is a high probability that the gondola will be overloaded if it is occupied by 12 more people, so it appears that the number of allowed passengers should be reduced.

19. a.   Select the **NORM.DIST** function. Enter the following values: x = 140, Mean = 165, Standard deviation = 45.6, Cumulative = TRUE. The function returns a probability of 0.2918. Select the **NORM.DIST** function again. Enter the following values: x = 211, Mean = 165, Standard deviation = 45.6, Cumulative = TRUE. The function returns a probability of 0.8435. $0.8435 - 0.2918 = 0.5517$

    b. $\mu_{\bar{x}} = \mu = 165.0$, $\sigma_{\bar{x}} = \sigma / \sqrt{n} = 45.6 / \sqrt{36} = 7.6$

         Select the **NORM.DIST** function. Click **OK**. Enter the following values: x = 140, Mean = 165, Standard deviation = 7.6, Cumulative = TRUE. Click **OK**. The function returns a probability of 0.0005. Select the **NORM.DIST** function again. Click **OK**. Enter the following values: x = 211, Mean = 165, Standard deviation = 7.6, Cumulative = TRUE. Click **OK**. The function returns a probability of 0.9999. $0.9990 - 0.0005 = 0.9994$

    c.   Part (a) because the ejection seats will be occupied by individual women, not groups of women.

21. a.   Select the **NORM.DIST** function. Enter the following values: x = 69.5, Mean = 72, Standard deviation = 2.4, Cumulative = TRUE. The function returns a probability of 0.1488. The probability the male passenger can fit through the doorway without bending is $1 - 0.1488 = 0.8512$.

    b. $\mu_{\bar{x}} = \mu = 69.5$, $\sigma_{\bar{x}} = \sigma / \sqrt{n} = 2.4 / \sqrt{100} = 0.24$

         Select the **NORM.DIST** function. Enter the following values: x = 69.5, Mean = 72, Standard deviation = 0.24, Cumulative = TRUE. The function returns a probability of 1.0406E–25 or 0+. The probability that the mean height of the 100 men is less than 72 in. is $1 - 0.0000 = 1.0000$ rounded to four decimal places.

c. The probability of part (a) is more relevant because it shows that 85.08% of male passengers will not need to bend. The result from part (b) gives us useful information about the comfort and safety of individual male passengers.

d. Because men are generally taller than women, a design that accommodates a suitable proportion of men will necessarily accommodate a greater proportion of women.

Section 6-5, Beyond the Basics

23. a. Yes. The sampling is without replacement and the sample size of 50 is greater than 5% of the finite population size of 275. $\sigma_{\bar{x}} = \frac{16}{\sqrt{50}} \sqrt{\frac{275-50}{275-1}} = 2.0504584$.

    b. Select the **NORM.DIST** function. Enter the following values: x = 95, Mean = 95.5, Standard deviation = 2.0504584, Cumulative = TRUE. The function returns a probability of 0.4037. Select the **NORM.DIST** function again. Enter the following values: x = 105, Mean = 95.5, Standard deviation = 2.0504584, Cumulative = TRUE. The function returns a probability of 1.0000 rounded to four decimal places. $1.0000 - 0.4037 = 0.5963$

25. a. $\mu = \frac{4+5+9}{3} = 6$, $\sigma = \sqrt{\frac{(4-6)^2 + (5-6)^2 + (9-6)^2}{3}} = 2.160246899$

    b. The possible samples of size 2 are:{ (4, 5), (4, 9), (5, 4), (5, 9), (9, 4), (9, 5)} which have the following means {4.5, 6.5, 4.5, 7, 6.5, 7} respectively.

    c. $\mu_{\bar{x}} = \frac{4.5+6.5+4.5+7+6.5+7}{6} = 6$ and

        $\sigma_{\bar{x}} = \sqrt{\frac{(4.5-6)^2 + (6.5-6)^2 + (4.5-6)^2 + (7-6)^2 + (6.5-6)^2 + (7-6)^2}{6}} = 1.08012345$

    d. It is clear that $\mu = \mu_{\bar{x}} = 6$. $\sigma_{\bar{x}} = \frac{2.160246899}{\sqrt{2}} \sqrt{\frac{3-2}{3-1}} = 1.08012345 = \sigma$

Section 6-6, Basic Skills and Concepts

1. The histogram should be approximately bell-shaped, and the normal quantile plot should have points that approximate a straight line pattern.

3. We must verify that the sample is from a population having a normal distribution. We can check for normality using a histogram, identifying the number of outliers, and constructing a normal quantile plot.

5. Not normal. The points show a systematic pattern that is not a straight line pattern.

7. Normal. The points are reasonably close to a straight line pattern, and there is no other pattern that is not a straight line pattern.

9. Open the **FLIGHTS** data file on your data disk. Select the **XLSTAT** add-in. Select **Visualizing data**. Select **Histograms**. Enter the **Arr Delay** data range inclusive of the label. Select **Discrete**. Select **Sample labels**. Click the **Charts** tab. Select **Histograms** and **Bars**. Select **Frequency** for ordinate of the histogram. Click **OK**. Not normal.

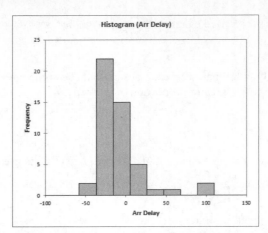

11. Open the **MBODY** data file on your data disk. Select the **XLSTAT** add-in. Select **Visualizing data**. Select **Histograms**. Enter the **SYS** data range inclusive of the label. Select **Discrete**. Select **Sample labels**. Click the **Charts** tab. Select **Histograms** and **Bars**. Select **Frequency** for ordinate of the histogram. Click **OK**. Normal.

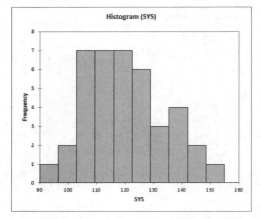

13. Open the **FLIGHTS** data file on your data disk. Select the **XLSTAT** add-in. Select **Describing data**. Select **Normality tests**. Enter the **Arr Delay** data range inclusive of the label. Select **Sample labels**. Click the **Charts** tab. Select **Normal Q-Q plots**. Click **OK**. Not normal.

Note that XLSTAT uses normalized scores rather than z-scores in the construction of a quantile plot.

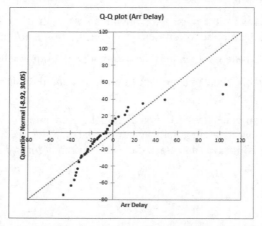

15. Open the **MBODY** data file on your data disk. Select the **XLSTAT** add-in. Select **Describing data**. Select **Normality tests**. Enter the **SYS** data range inclusive of the label. Select **Sample labels**. Click the **Charts** tab. Select **Normal Q-Q plots**. Click **OK**. Normal.

Note that XLSTAT uses normalized scores rather than z-scores in the construction of a quantile plot.

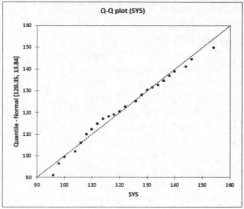

17. Enter the label **Feet** in cell A1 of an Excel worksheet followed by the five data values. Sort the data values from lowest to highest. With a sample size $n = 5$, each value represents a proportion of 1/5 of the sample. Next, you identify the cumulative areas to the left of the corresponding sample values. The formulas are:

$\dfrac{1}{2n}, \dfrac{3}{2n}, \dfrac{5}{2n}, \dfrac{7}{2n}, \dfrac{9}{2n}$. The specific areas, then, are: $\dfrac{1}{10}, \dfrac{3}{10}, \dfrac{5}{10}, \dfrac{7}{10}, \dfrac{9}{10}$. Expressed as decimals, the specific areas

are: 0.1, 0.3, 0.5, 0.7, and 0.9. Enter the label **z score** in cell B1. You will now use Excel's NORM.S.INV function to find the corresponding z scores. Click in cell **B2** where the z score corresponding to an area of 0.1 will be placed. Click **Formulas** at the top of the screen. Select **Insert Function**. Select the **Statistical** category. Select the **NORM.S.INV** function. Click **OK**. Enter the following value: Probability = 0.1. Click **OK**. The function returns $z = -1.28$. Continue in the same way to find the remaining z scores. The feet and z-scores are displayed below in the table on the left.

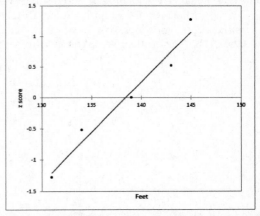

	A	B
1	Feet	z score
2	131	-1.2816
3	134	-0.5244
4	139	0
5	143	0.5244
6	145	1.28155

The feet and z score pairs are (x,y) coordinates that you will plot in a scatter plot. Select the XLSTAT add-in. Select **Visualizing data**. Select **Scatter plots**. Feet is the x variable.

Enter the data range of the Feet variable inclusive of the label. z score is the y variable. Enter the data range of the z score variable inclusive of the label. Select **Variable labels**. Click **OK**. Right-click directly on one of the plotted points in the graph. Select **Add trendline** from the menu that appears. Select **Linear** from trendline options. The normal quantile plot is displayed on the right. The data appear to be from a population with a normal distribution.

19. Enter the label **cm^3** in cell A1 of an Excel worksheet followed by the eight data values. Sort the data values from lowest to highest. With a sample size $n = 8$, each value represents a proportion of 1/8 of the sample. Next, you identify the cumulative areas to the left of the corresponding sample values. The formulas are:

$\dfrac{1}{2n}, \dfrac{3}{2n}, \dfrac{5}{2n}, \dfrac{7}{2n}, \dfrac{9}{2n}, \dfrac{11}{2n}, \dfrac{13}{2n}, \dfrac{15}{2n}$. The specific areas, then, are: $\dfrac{1}{16}, \dfrac{3}{16}, \dfrac{5}{16}, \dfrac{7}{16}, \dfrac{9}{16}, \dfrac{11}{16}, \dfrac{13}{16}, \dfrac{15}{16}$. Expressed as

decimals, the specific areas are: 0.0625, 0.1875, 0.3125, 0.4375, 0.5625, 0.6875, 0.8125, and 0.9375 Enter the label **z score** in cell B1. You will now use Excel's NORM.S.INV function to find the corresponding z scores. Click in cell **B2** where the z score corresponding to an area of 0.0625 will be placed. Click **Formulas** at the top of the screen. Select **Insert Function**. Select the **Statistical** category. Select the **NORM.S.INV** function. Enter the following value: Probability = 0.0625. The function returns $z = -1.53$. Continue in the same way to find the remaining z scores. The cm^3 and z-scores are displayed on the next page in the table on the left.

	A	B
1	cm^3	z score
2	1034	-1.5341
3	1051	-0.8871
4	1067	-0.4888
5	1070	-0.1573
6	1079	0.15731
7	1079	0.48878
8	1173	0.88715
9	1272	1.53412

The cm^3 and z score pairs are (x,y) coordinates that you will plot in a scatter plot. Select the XLSTAT add-in. Select **Visualizing data**. Select **Scatter plots**. Cm^3 is the x variable. Enter the data range of the cm^3 variable inclusive of the label. z score is the y variable. Enter the data range of the z score variable inclusive of the label. Select **Variable labels**. Click **OK**. Right-click directly on one of the plotted points in the graph. Select **Add trendline** from the menu that appears. Select **Linear** from trendline options. The normal quantile plot is displayed on the right. The data appear to be from a population with a distribution that is not normal.

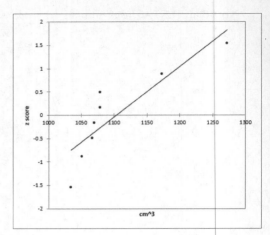

Section 6-6, Beyond the Basics

21. a. Yes

    b. Yes

    c. No

23. Enter the label **Thousands of dollars** in cell A1 of an Excel worksheet followed by the data given in Exercise 23. Select the **XLSTAT** add-in. Select **Describing data**. Select **Normality tests**. Enter the **Thousands of dollars** data range inclusive of the label. Select **Sample labels**. Click the **Charts** tab. Select **Normal Q-Q plots**. Click **OK**. The original values are not from a normally distributed population.

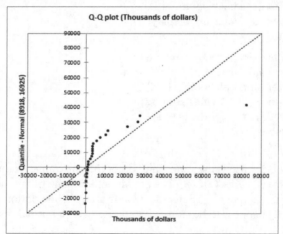

Q-Q plot (Thousands of dollars)

Enter the label **Logarithm** in cell B1 of the worksheet. Then use Excel's **LOG10** function to take the logarithm of each value. Use XLSTAT to construct a normality Q-Q plot of the logarithms.

After taking the logarithm of each value, the values appear to be from a normally distributed population. The original values are from a population with a lognormal distribution.

Note that XLSTAT uses normalized scores rather than z-scores in the construction of a quantile plot.

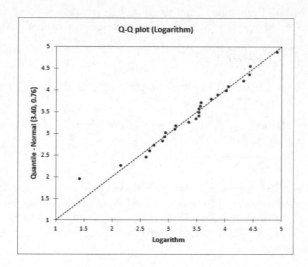

Section 6-7, Basic Skills and Concepts

1. The Minitab display shows that the region representing 235 wins is a rectangle. The result of 0.0068 is an approximation, but the result of 0.0066 is better because it is based on an exact calculation. The approximation differs from the exact result by a very small amount.

3. $p = \dfrac{1}{5} = 0.2$, $q = \dfrac{4}{5} = 0.8$, $\mu = 25 \cdot 0.2 = 5$, $\sigma = \sqrt{25 \cdot 0.2 \cdot 0.8} = 2$. The value of 5 for the mean shows that for many people who make random guesses for the 25 questions, the mean number of correct answers is 5. For many people who make random guesses, the standard deviation of 2 is a measure of how much the numbers of correct responses vary.

5. The requirements are satisfied with a mean of $13 \cdot 0.4 = 5.2$ and the standard deviation of $\sqrt{13 \cdot 0.4 \cdot 0.6} = 1.7663$. Select Excel's **NORM.DIST** function. Enter the following values: x = 2.5, Mean = 5.2, Standard deviation = 1.7663, Cumulative = TRUE. The function returns a probability of 0.0632.

7. The requirement of $nq \geq 5$ is not satisfied. Normal approximation should not be used.

For Exercises 9-11: $\mu = 100 \cdot 0.22 = 22$, $\sigma = \sqrt{100 \cdot 0.22 \cdot 0.78} = 4.1425$

9. Select Excel's **NORM.DIST** function. Enter the following values: x = 19.5, Mean = 22, Standard deviation = 4.1425, Cumulative = TRUE. The function returns a probability of 0.2731.

11. Select Excel's **NORM.DIST** function. Enter the following values: x = 22.5, Mean = 22, Standard deviation = 4.1425, Cumulative = TRUE. The function returns a probability of 0.5480. Select Excel's **NORM.DIST** function again. Enter the following values: x = 23.5, Mean = 22, Standard deviation = 4.1425, Cumulative = TRUE. The function returns a probability of 0.6414. $0.6414 - 0.5480 = 0.0934$

13. a. $\mu = 611 \cdot 0.3 = 183.3$, $\sigma = \sqrt{611 \cdot 0.3 \cdot 0.7} = 11.3274$

   Select Excel's **NORM.DIST** function. Enter the following values: x = 171.5, Mean = 183.3, Standard deviation = 11.3274, Cumulative = TRUE. The function returns a probability of 0.1488. Select Excel's **NORM.DIST** function again. Enter the following values: x = 172.5, Mean = 183.3, Standard deviation = 11.3274, Cumulative = TRUE. The function returns a probability of 0.1702. Using normal approximation: $0.1702 - 0.1488 = 0.0214$.

   To use the binomial, select Excel's **BINOM.DIST** function. Enter the following values: Number of successes = 172, Trials = 611, Probability = 0.3, Cumulative = FALSE. The function returns 0.0217.

b.   Select Excel's **NORM.DIST** function. Enter the following values: x = 172.5, Mean = 183.3, Standard deviation = 11.3274, Cumulative = TRUE. The function returns a probability of 0.1702.

To use the binomial, select Excel's **BINOM.DIST** function. Enter the following values: Number of successes = 172, Trials = 611, Probability = 0.3, Cumulative = TRUE. The function returns 0.1703.

c.   The result from part (b) is useful. We want the probability of getting a result that is at least as extreme as the one obtained.

d.   If the 30% rate is correct, there is a good chance (17%) of getting 172 or fewer calls overturned, so there is not strong evidence against the 30% rate.

15. a. $\mu = 580 \cdot 0.75 = 435$, $\sigma = \sqrt{580 \cdot 0.75 \cdot 0.25} = 10.4283$

Select Excel's **NORM.DIST** function. Enter the following values: x = 427.5, Mean = 435, Standard deviation = 10.4283, Cumulative = TRUE. The function returns a probability of 0.2360. Select Excel's **NORM.DIST** function again. Enter the following values: x = 428.5, Mean = 435, Standard deviation = 10.4283, Cumulative = TRUE. The function returns a probability of 0.2665. Using normal approximation: $0.2665 - 0.2360 = 0.0305$.

To use the binomial, select Excel's **BINOM.DIST** function. Enter the following values: Number of successes = 428, Trials = 580, Probability = 0.75, Cumulative = FALSE. The function returns 0.0301.

b.   Select Excel's **NORM.DIST** function. Enter the following values: x = 428.5, Mean = 435, Standard deviation = 10.4283, Cumulative = TRUE. The function returns a probability of 0.2665.

To use the binomial, select Excel's **BINOM.DIST** function. Enter the following values: Number of successes = 428, Trials = 580, Probability = 0.75, Cumulative = TRUE. The function returns 0.2650.

The result of 428 peas with green pods is not unusually low.

c.   The result from part (b) is useful. We want the probability of getting a result that is at least as extreme as the one obtained.

d.   No. Assuming that Mendel's probability of 3/4 is correct, there is a good chance (26.7%) of getting the results that were obtained. The obtained results do not provide strong evidence against the claim that the probability of a pea having a green pod is ¾.

17. a. $\mu = 945 \cdot 0.5 = 472.5$, $\sigma = \sqrt{945 \cdot 0.5 \cdot 0.5} = 15.3704$

Select Excel's **NORM.DIST** function. Enter the following values: x = 878.5, Mean = 472.5, Standard deviation = 15.3704, Cumulative = TRUE. The function returns a probability of 1.0000. Select Excel's **NORM.DIST** function again. Enter the following values: x = 879.5, Mean = 472.5, Standard deviation = 15.3704, Cumulative = TRUE. The function returns a probability of 1.0000. Using normal approximation: $1.0000 - 1.0000 = 0.0000$, which is a very small probability that is extremely close to 0.

To use the binomial, select Excel's **BINOM.DIST** function. Enter the following values: Number of successes = 879, Trials = 945, Probability = 0.5, Cumulative = FALSE. The function returns 1.4451E–182, a probability very close to 0.

b.   Select Excel's **NORM.DIST** function. Enter the following values: x = 878.5, Mean = 472.5, Standard deviation = 15.3704, Cumulative = TRUE. The function returns a probability of 1.0000. $1.0000 - 1.0000 = 0.0000$, which is a very small probability that is extremely close to 0.

To use the binomial, select Excel's **BINOM.DIST** function. Enter the following values: Number of successes = 879, Trials = 945, Probability = 0.5, Cumulative = TRUE. The function returns 1.0000. $1.0000 - 1.0000 = 0.0000$, which is a very small probability that is extremely close to 0.

If boys and girls are equally likely, 879 girls in 945 births is unusually high.

c.   The result from part (b) is more relevant, because we want the probability of a result that is at least as extreme as the one obtained.

d.   Yes. It is very highly unlikely that we would get 879 girls in 945 births by chance. Given that the 945 couples were treated with the XSORT method, it appears that this method is effective in increasing the likelihood that a baby will be a girl.

19. $\mu = 1002 \cdot 0.61 = 611.22$, $\sigma = \sqrt{1002 \cdot 0.61 \cdot 0.39} = 15.4394$

Select Excel's **NORM.DIST** function. Enter the following values: x = 700.5, Mean = 611.22, Standard deviation = 15.4394, Cumulative = TRUE. The function returns a probability of 1.0000, rounded to four decimal places. 1.0000−1.0000=0.0000, which is a very small probability that is extremely close to 0.

To use the binomial, select Excel's **BINOM.DIST** function. Enter the following values: Number of successes = 701, Trials = 1002, Probability = 0.61, Cumulative = TRUE. The function returns 1.0000, rounded to four decimal places. 1.0000−1.0000=0.0000, which is a very small probability that is extremely close to 0.

The result suggests that the surveyed people did not respond accurately.

21. $\mu = 50 \cdot 0.2 = 10$, $\sigma = \sqrt{50 \cdot 0.2 \cdot 0.8} = 2.8284$

Select Excel's **NORM.DIST** function. Enter the following values: x = 6.5, Mean = 10, Standard deviation = 2.8284, Cumulative = TRUE. The function returns a probability of 0.1080.

To use the binomial, select Excel's **BINOM.DIST** function. Enter the following values: Number of successes = 6, Trials = 50, Probability = 0.2, Cumulative = TRUE. The function returns 0.1034.

Because that probability is not very small, the evidence against the rate of 20% is not very strong.

23. The probability of 170 or fewer should be computed. $\mu = 1000 \cdot 0.20 = 200$, $\sigma = \sqrt{1000 \cdot 0.2 \cdot 0.8} = 12.6491$

Select Excel's **NORM.DIST** function. Enter the following values: x = 169.5, Mean = 200, Standard deviation = 12.6491, Cumulative = TRUE. The function returns a probability of 0.0079.

To use the binomial, select Excel's **BINOM.DIST** function. Enter the following values: Number of successes = 170, Trials = 1000, Probability = 0.2, Cumulative = TRUE. The function returns a probability of 0.0089.

Because the probability of 170 or fewer is so small with the assumed 20% rate, it appears that the rate is actually less than 20%.

Section 6-7, Beyond the Basics

25. a.   In order to make a profit Marc will need to win over $1000. With 35:1 odds a $5 bet wins $175. Therefore, Marc needs 6 winning bets in order to make a profit. $\mu = 200 \cdot \frac{1}{38} = 5.2632$, $\sigma = \sqrt{200 \cdot \frac{1}{38} \cdot \frac{37}{38}} = 2.2638$

$z_{x=5.5} = \dfrac{5.5 - 5.2632}{2.2638} = 0.10$ which has a probability of 1.0000−0.5398=0.4602 to the right of it.

(Excel using normal approximation: 0.4583; Excel using binomial: 0.4307)

b.   Since the odds of winning are 1:1 Marc would need 101 wins or more to make a profit. $\mu = 200 \cdot \frac{244}{495} = 98.5859$, $\sigma = \sqrt{200 \cdot \frac{244}{495} \cdot \frac{251}{495}} = 7.0704$

$z_{x=100.5} = \dfrac{100.5 - 98.5859}{7.0704} = 0.27$ which has a probability of 1.0000−0.6064=0.3936

(Excel using normal approximation: 0.3933; Excel using binomial: 0.3932)

c.   The roulette game provides a better likelihood of making a profit

Chapter Quick Quiz

1. $\mu = 0$ and $\sigma = 1$

2.

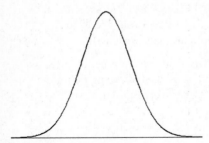

3.   You will use Excel's NORM.S.INV function to solve the problem. Click **Formulas** at the top of the screen. Select **Insert Function**. Select the **Statistical** category. Select the **NORM.S.INV** function. Enter the following value: Probability = 0.98. The function returns 2.0537. $z = 2.0537$

4.   You will use Excel's NORM.S.DIST function to solve the problem. Click **Formulas** at the top of the screen. Select **Insert Function**. Select the **Statistical** category. Select the **NORM.S.DIST** function. Enter the following values: z = −1, Cumulative = TRUE. The function returns 0.1587. $P(z < -1) = 1 - 0.1587 = 0.8413$.

5.   You will use Excel's NORM.S.DIST function to solve the problem. Click **Formulas** at the top of the screen. Select **Insert Function**. Select the **Statistical** category. Select the **NORM.S.DIST** function. Enter the following values: z = 1.37, Cumulative = TRUE. The function returns 0.9147. Select the **NORM.S.DIST** function again. Enter the following values: z = 2.42, Cumulative = TRUE. The function returns 0.9922.
$P(1.37 < z < 2.42) = P(z < 2.42) - P(z < 1.37) = 0.9922 - 0.9147 = 0.0775$

6.   Select Excel's **NORM.DIST** function. Enter the following values: x = 4.2, Mean = 4.577, Standard deviation = 0.382, Cumulative = TRUE. The function returns a probability of 0.1618.

7.   Select Excel's **NORM.DIST** function. Enter the following values: x = 5.4, Mean = 4.577, Standard deviation = 0.382, Cumulative = TRUE. The function returns a probability of 0.9844. The probability that a randomly selected woman has a red blood cell count above the specified normal range is $1 - 0.9844 = 0.0156$.

8.   Select the **NORM.INV** function. Enter the following values: Probability = 0.8, Mean = 4.577, Standard deviation = .382. The function returns a red blood cell count of 4.898.

9. $\mu_{\bar{x}} = \mu = 4.577$, $\sigma_{\bar{x}} = \sigma / \sqrt{n} = 0.382 / \sqrt{25} = 0.0764$

   Select the **NORM.DIST** function. Enter the following values: x = 4.444, Mean = 4.577, Standard deviation = 0.0764, Cumulative = TRUE. The function returns a probability of 0.0409.

10. Select Excel's **NORM.DIST** function. Enter the following values: x = 4.2, Mean = 4.577, Standard deviation = 0.382, Cumulative = TRUE. The function returns a probability of 0.1618. Select Excel's **NORM.DIST** function again. Enter the following values: x = 5.4, Mean = 4.577, Standard deviation = 0.382, Cumulative = TRUE. The function returns a probability of 0.9844. The percentage in the normal range from 4.2 to 5.4 $= 0.9844 - 0.1618 = 0.8226$ or 82.26%.

Review Exercises

1.   a.   Select the **NORM.S.DIST** function. Enter the following values: z = 2.93, Cumulative = TRUE. The function returns 0.9147. Select the **NORM.S.DIST** function again. Enter the following values: z = 2.42, Cumulative = TRUE. The function returns 0.9983.

     b.   Select the **NORM.S.DIST** function. Enter the following values: z = −1.53, Cumulative = TRUE. The function returns 0.0630. The probability of a bone density test greater than −1.53 $= 1 - 0.0630 = 0.9370$.

c. Select the **NORM.S.DIST** function. Enter the following values: $z = -1.07$, Cumulative = TRUE. The function returns 0.1423. Select the **NORM.S.DIST** function again. Enter the following values: 2.07, Cumulative = TRUE. The function returns 0.9808. $0.9808 - 0.1423 = 0.8385$.

d. Select the **NORM.S.INV** function. Enter the following value: Probability = .3. The function returns –0.52.

e. $\sigma_{\bar{x}} = \sigma / \sqrt{n} = 1/16 = 0.25$

Select the **NORM.DIST** function. Enter the following values: x = 0.27, Mean = 0, Standard deviation = 0.25, Cumulative = TRUE. The function returns 0.8599. $1 - 0.8599 = 0.1401$

2. a. Select the **NORM.DIST** function. Enter the following values: x = 1605, Mean = 1516, Standard deviation = 63, Cumulative = TRUE. The function returns 0.9211. $1 - 0.9211 = 0.0789$; 7.89% of women will find that height uncomfortable.

b. Select the **NORM.INV** function. Enter the following values: Probability = 0.01, Mean = 1516, Standard deviation = 63. The function returns 1369.4 mm.

3. a. Select the **NORM.DIST** function. Enter the following values: x = 1500, Mean = 1634, Standard deviation = 66, Cumulative = TRUE. The function returns 0.0212. $1 - 0.0212 = 0.9788$; 97.88% of men will find that height uncomfortable.

b. Select the **NORM.INV** function. Enter the following values: Probability = 0.95, Mean = 1634, Standard deviation = 66. The function returns 1742.6 mm.

4. a. Normal distribution

b. $\mu_{\bar{x}} = 21.1$

c. $\sigma_{\bar{x}} = \dfrac{5.1}{\sqrt{80}} = 0.57$

5. a. An unbiased estimator is a statistic that targets the value of the population parameter in the sense that the sampling distribution of the statistic has a mean that is equal to the mean of the corresponding parameter.

b. Mean, variance and proportion

c. True

6. a. Select the **NORM.DIST** function. Enter the following values: x = 72, Mean = 69.5, Standard deviation = 2.4, Cumulative = TRUE. The function returns 0.8512. 85.12% of men will fit without bending. With about 15% of all men needing to bend, the design does not appear to be adequate, but the Mark VI monorail appears to be working quite well in practice.

b. Select the **NORM.INV** function. Enter the following values: Probability = 0.99, Mean = 69.5, Standard deviation = 2.4. The function returns 75.1 in.

7. a. Select the **NORM.DIST** function. Enter the following values: x = 175, Mean = 182.9, Standard deviation = 40.9, Cumulative = TRUE. The function returns 0.4234. $1 - 0.4234 = 0.5766$

b. $\sigma_{\bar{x}} = \dfrac{40.9}{\sqrt{213}} = 2.8024$

Select the **NORM.DIST** function. Enter the following values: x = 175, Mean = 182.9, Standard deviation = 2.8024, Cumulative = TRUE. The function returns 0.0024. $1 - 0.0024 = 0.9976$; yes, if the plane is full of male passengers, it is highly likely that it is overweight.

8. a. Enter the label **Thousands of dollars** in cell A1 of an Excel worksheet followed by the data given in Exercise 8. Select the **XLSTAT** add-in. Select **Describing data**. Select **Histograms**. Enter the range of the salary data inclusive of the label. Select **Sample labels**. Click the **Charts** tab. Select **Histograms** and **Bars**. Select **Frequency** for the ordinate of the histogram. Click **OK**.

Next, construct a normal quantile plot. Select **Describing data**. Select **Normality tests**. Enter the range of the salary data inclusive of the label. Select **Sample labels**. Click the **Charts** tab. Select **Normal Q-Q plots**. Click **OK**.

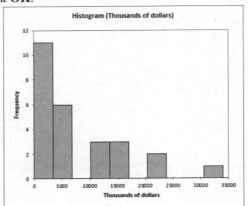

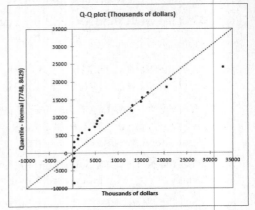

No. A histogram is far from bell-shaped. A normal quantile plot reveals a pattern of points that is far from a straight-line pattern. Note that XLSTAT uses normalized scores rather than z-scores in the construction of a quantile plot.

b.   No. The sample has a size of 26 which does not satisfy the condition at least 30, and the values do not appear to be from a population having a normal distribution.

9. $\mu = 1064 \cdot 0.75 = 798$, $\sigma = \sqrt{1064 \cdot 0.75 \cdot .25} = 14.1244$

Select Excel's **NORM.DIST** function. Enter the following values: x = 787.5, Mean = 798, Standard deviation = 14.1244, Cumulative = TRUE. The function returns a probability of 0.2286.

To use the binomial, select Excel's **BINOM.DIST** function. Enter the following values: Number of successes = 787, Trials = 1064, Probability = 0.75, Cumulative = TRUE. The function returns 0.2278.

The occurrence of 787 offspring plants with long stem is not unusually low because its probability is not small. The results are consistent with Mendel's claimed proportion of 3/4

10. a. $\mu = 64 \cdot 0.8 = 51.2$, $\sigma = \sqrt{64 \cdot 0.8 \cdot 0.2} = 3.2$

Select Excel's **NORM.DIST** function. Enter the following values: x = 50.5, Mean = 51.2, Standard deviation = 3.2, Cumulative = TRUE. The function returns a probability of 0.4134. 1−0.4134=0.5866

To use the binomial, select Excel's **BINOM.DIST** function. Enter the following values: Number of successes = 50, Trials = 64, Probability = 0.8, Cumulative = TRUE. The function returns 0.4019. 1−0.4019=0.5981

b.   Select Excel's **NORM.DIST** function. Enter the following values: x = 50.5, Mean = 51.2, Standard deviation = 3.2, Cumulative = TRUE. The function returns a probability of 0.4134.

Select Excel's **NORM.DIST** function again. Enter the following values: x = 49.5, Mean = 51.2, Standard deviation = 3.2, Cumulative = TRUE. The function returns a probability of 0.2976.

The probability of exactly 3 = 0.4134−0.2976=0.1158 .

To use the binomial, select Excel's **BINOM.DIST** function. Number of successes = 50, Trials = 64, Probability = 0.8, Cumulative = FALSE. The function returns 0.1119.

Cumulative Review Exercises

1. Enter the label **Thousands of dollars** in cell A1 of an Excel worksheet followed by the annual salary data given in Exercise 1. Click on **DATA** at the top of the screen. Select **Data Analysis** and select **Descriptive Statistics**. Enter the input range of the salary data, including the label (**A1:A6**). Select **Labels in First Row** and select **Summary statistics**. Click **OK**.

Thousands of dollars	
Mean	10300
Standard Error	2482.942
Median	14000
Mode	14500
Standard Deviation	5552.027
Sample Variance	30825000
Kurtosis	-3.08391
Skewness	-0.64391
Range	11000
Minimum	3500
Maximum	14500
Sum	51500
Count	5

   a. Expressed in dollars, $\bar{x} = 10300 \times 1,000 = \$10,300,000$.

   b. Expressed in dollars, the median $= 14000 \times 1000 = \$14,000,000$.

   c. Expressed in dollars, $s = 5552.027 \times 1000 = \$5,552,027$.

   d. Expressed in dollars,
$s^2 = 30825000 \times 1000 = \$30,825,003,810,000$ square dollars.

   e. Select Excel's **STANDARDIZE** function. Enter the following values: x = 14500000, Mean = 10300000, Standard deviation = 5552027. The function returns $z = 0.7564$.

   f. Ratio

   g. Discrete

   h. No, the starting players are likely to be the best players who receive the highest salaries.

2. a. $\bar{A}$ is the event of selecting someone who does not have the belief that college is not a good investment. NOTE: This is not the same as selecting someone who believes that college is a good investment.

   b. $P(\bar{A}) = 1 - 0.1 = 0.9$

   c. $P = 0.1 \cdot 0.1 \cdot 0.1 = 0.001$

   d. The sample is a voluntary response sample. This suggests that the 10% rate might not be very accurate, because people with strong feelings or interest about the topic are more likely to respond.

3. a. Select Excel's **NORM.DIST** function. Enter the following values: x = 2500, Mean = 3369, Standard deviation = 567, Cumulative = TRUE. The function returns a probability of 0.0627.

   b. Select Excel's **NORM.INV** function. Enter the following values: Probability = .1, Mean = 3369, Standard deviation = 567. The function returns 2642.4 g.

   c. Select Excel's **NORM.DIST** function. Enter the following values: x = 1500, Mean = 3369, Standard deviation = 567, Cumulative = TRUE. The function returns a probability of 0.00049.

   d. $\sigma_{\bar{x}} = \dfrac{567}{\sqrt{25}} = 113.4$

   Select Excel's **NORM.DIST** function. Enter the following values: x = 3400, Mean = 3369, Standard deviation = 113.4, Cumulative = TRUE. The function returns a probability of 0.6077. $1 - 0.6077 = 0.3923$

4. a. The vertical scale does not start at 0, so differences are somewhat distorted. By using a scale ranging from 1 to 29 for frequencies that range 2 to 14, the graph is flattened, so differences are not shown as they should be.

   b. The graph depicts a distribution that is not exactly normal, but it is approximately normal because it is roughly bell shaped.

   c. Minimum: 42 years; maximum: 70 years. Using the range rule of thumb, the standard deviation is estimated to be $\dfrac{70 - 42}{4} = 7$ years. The estimate of 7 years is very close to the actual standard deviation of 6.6 years, so the range rule of thumb works quite well here.

5.   a. $P(X = 3) = 0.1 \cdot 0.1 \cdot 0.1 = 0.001$

    b. $P(X \geq 1) = 1 - P(X = 0) = 1 - (0.9 \cdot 0.9 \cdot 0.9) = 0.271$

    c.   The requirement that $np \geq 5$ is not satisfied, indicating that the normal approximation would result in errors that are too large.

    d. $\mu = 50 \cdot 0.1 = 5$

    e. $\sigma = \sqrt{50 \cdot 0.1 \cdot 0.9} = 2.1213$

    f.   No, 8 is within two standard deviations of the mean and is within the range of values that could easily occur by chance.

Chapter 7

Estimates and Sample Sizes

Section 7-2, Basic Skills and Concepts

1. The confidence level (such as 95%) was not provided.

3. $\hat{p} = 0.26$ is the sample proportion; $\hat{q} = 0.74$ (found from evaluating $1 - \hat{p}$); $n = 1910$ is the sample size; $E = 0.03$ is the margin of error; p is the population proportion, which is unknown. The value of α is 0.05.

For Exercises 5-7, you will use Excel's NORM.S.INV function.

5. $\alpha/2 = 0.20/2 = 0.10$. For the upper 0.10, NORM.S.INV(0.90) = 1.28.

7. $\alpha/2 = 0.10/2 = 0.05$. For the upper 0.05, NORM.S.INV(0.95) = 1.645

9. $E = \dfrac{0.186 - 0.0641}{2} = 0.061$, so 0.125 ± 0.061

11. $0.0268 < p < 0.133$

13. a. $\hat{p} = \dfrac{531}{1002} = 0.530$

    b. $\alpha/2 = 0.05/2 = 0.025$. For the upper 0.975, NORM.S.INV(0.975) = 1.96

    $$E = z_{\alpha/2}\sqrt{\dfrac{\hat{p}\hat{q}}{n}} = 1.96\sqrt{\dfrac{\left(\frac{531}{1002}\right)\left(\frac{471}{1002}\right)}{1002}} = 0.0309$$

    c. $\hat{p} - E < p < \hat{p} - E \Rightarrow 0.530 - 0.0309 < p < 0.530 - 0.0309 \Rightarrow 0.499 < p < 0.561$

    d. We have 95% confidence that the interval from 0.499 to 0.561 actually does contain the true value of the population proportion.

15. a. $\hat{p} = \dfrac{1083}{2518} = 0.430$

    b. $\alpha/2 = 0.10/2 = 0.05$. For the upper 0.05, NORM.S.INV(0.95) = 1.65

    $$E = z_{\alpha/2}\sqrt{\dfrac{\hat{p}\hat{q}}{n}} = 1.65\sqrt{\dfrac{\left(\frac{1083}{2518}\right)\left(\frac{1435}{2518}\right)}{2518}} = 0.0162$$

    $$\hat{p} - E < p < \hat{p} - E$$

    c. $0.430 - 0.0162 < p < 0.430 - 0.0162$

    $$0.414 < p < 0.446$$

    d. We have 90% confidence that the interval from 0.414 to 0.446 actually does contain the true value of the population proportion.

17. a. $\hat{p} = \dfrac{879}{945} = 0.930$

    b. $\alpha/2 = 0.05/2 = 0.025$. For the upper 0.975, NORM.S.INV(0.975) = 1.96

    $$\hat{p} \pm z_{\alpha/2}\sqrt{\dfrac{\hat{p}\hat{q}}{n}} = \dfrac{879}{945} \pm 1.96\sqrt{\dfrac{\left(\frac{879}{945}\right)\left(\frac{66}{945}\right)}{945}}$$

    $$0.914 < p < 0.946$$

    c.   Yes. The true proportion of girls with the XSORT method is substantially greater than the proportion of (about) 0.5 that is expected when no method of gender selection is used.

19. a.   0.5

    b. $\hat{p} = \dfrac{123}{280} = 0.439$

    c. $\hat{p} \pm z_{\alpha/2}\sqrt{\dfrac{\hat{p}\hat{q}}{n}} = \dfrac{123}{280} \pm 2.576\sqrt{\dfrac{\left(\frac{123}{280}\right)\left(\frac{157}{280}\right)}{280}}$

       $0.363 < p < 0.515$ or $36.3\% < p < 51.5\%$

    d.   If the touch therapists really had an ability to select the correct hand by sensing an energy field, their success rate would be significantly greater than 0.5, but the sample success rate of 0.439 and the confidence interval suggest that they do not have the ability to select the correct hand by sensing an energy field.

21. a. $427(0.29) = 124$

    b. $\alpha/2 = 0.05/2 = 0.025$. For the upper 0.975, NORM.S.INV(0.975) = 1.96

       $\hat{p} \pm z_{\alpha/2}\sqrt{\dfrac{\hat{p}\hat{q}}{n}} = 0.29 \pm z_{0.025}\sqrt{\dfrac{(0.29)(0.71)}{427}}$

       $0.247 < p < 0.333$ or $24.7\% < p < 33.3\%$

    c.   Yes. Because all values of the confidence interval are less than 0.5, the confidence interval shows that the percentage of women who purchase books online is very likely less than 50%.

    d.   No. The confidence interval shows that it is possible that the percentage of women who purchase books online could be less than 25%.

    e.   Nothing.

23. a. $514(0.459) = 236$

    b. $\alpha/2 = 0.05/2 = 0.025$. For the upper 0.975, NORM.S.INV(0.975) = 1.96

       $\hat{p} \pm z_{\alpha/2}\sqrt{\dfrac{\hat{p}\hat{q}}{n}} = 0.459 \pm z_{0.025}\sqrt{\dfrac{(0.459)(0.541)}{514}}$ (using $x = 236$: $0.403 < p < 0.516$).

           $0.402 < p < 0.516$

    c. $\hat{p} \pm z_{\alpha/2}\sqrt{\dfrac{\hat{p}\hat{q}}{n}} = 0.459 \pm z_{0.10}\sqrt{\dfrac{(0.459)(0.541)}{514}}$

           $0.431 < p < 0.487$

    d.   The 95% confidence interval is wider than the 80% confidence interval. A confidence interval must be wider in order to be more confident that it captures the true value of the population proportion. (See Exercise 4.)

25. $\alpha/2 = 0.02/2 = 0.01$. For the upper 0.01, NORM.S.INV(0.99) = 2.3263

    $\hat{p} \pm z_{\alpha/2}\sqrt{\dfrac{\hat{p}\hat{q}}{n}} = 0.08 \pm z_{0.01}\sqrt{\dfrac{(0.08)(0.92)}{100}} = 0.08 \pm 0.0631$

    $0.0169 < p < 0.1431$

27. a. $\alpha/2 = 0.10/2 = 0.05$. For the upper 0.05, NORM.S.INV(0.95) = 1.645

       $\hat{p} \pm z_{\alpha/2}\sqrt{\dfrac{\hat{p}\hat{q}}{n}} = 0.000321 \pm z_{0.05}\sqrt{\dfrac{(0.000321)(0.999679)}{420,095}} = 0.000321 \pm 0.0000454648$

           $0.0276\% < p < 0.0366\%$

(using $x = 135$: $0.0276\% < p < 0.0367\%$).

b.   No, because 0.0340% is included in the confidence interval.

29. $n = \dfrac{[z_{\alpha/2}]^2 \, \hat{p}\hat{q}}{E^2} = \dfrac{[1.645]^2 \, (0.25)}{0.03^2} = 752$

31. $n = \dfrac{[z_{\alpha/2}]^2 \, \hat{p}\hat{q}}{E^2} = \dfrac{[2.575]^2 \, (0.15)(0.85)}{0.05^2} = 339$

33. a. $n = \dfrac{[z_{\alpha/2}]^2 \, \hat{p}\hat{q}}{E^2} = \dfrac{[1.96]^2 \, (0.25)}{0.025^2} = 1537$

b. $n = \dfrac{[z_{\alpha/2}]^2 \, \hat{p}\hat{q}}{E^2} = \dfrac{[1.96]^2 \, (0.38)(0.62)}{0.025^2} = 1449$

35. a. $\alpha/2 = 0.10/2 = 0.05$. For the upper 0.05, NORM.S.INV(0.95) = 1.645

$n = \dfrac{[z_{\alpha/2}]^2 \, \hat{p}\hat{q}}{E^2} = \dfrac{[1.645]^2 \, (0.25)}{0.05^2} = 271$

b. $n = \dfrac{[z_{\alpha/2}]^2 \, \hat{p}\hat{q}}{E^2} = \dfrac{[1.645]^2 \, (0.85)(0.15)}{0.05^2} = 138$

c.   No. A sample of students at the nearest college is a convenience sample, not a simple random sample, so it is very possible that the results would not be representative of the population of adults.

37.   Greater height does not appear to be an advantage for presidential candidates. If greater height is an advantage, then taller candidates should win substantially more than 50% of the elections, but the confidence interval shows that the percentage of elections won by taller candidates is likely to be anywhere between 36.2% and 69.7%.

$\hat{p} = \dfrac{18}{34} = 0.529$. $\hat{p} \pm z_{\alpha/2}\sqrt{\dfrac{\hat{p}\hat{q}}{n}} = \dfrac{18}{34} \pm 1.96\sqrt{\dfrac{\left(\frac{18}{34}\right)\left(\frac{16}{34}\right)}{34}}$

$$0.362 < p < 0.697 \text{ or } 36.2\% < p < 69.7\%.$$

Section 7-2, Beyond the Basics

39. a. $n = \dfrac{N\hat{p}\hat{q}[z_{\alpha/2}]^2}{\hat{p}\hat{q}[z_{\alpha/2}]^2 + (N-1)E^2} = \dfrac{200(0.5)(0.5)[1.96]^2}{(0.5)(0.5)[1.96]^2 + (200-1)0.025^2} = 178$

b. $n = \dfrac{N\hat{p}\hat{q}[z_{\alpha/2}]^2}{\hat{p}\hat{q}[z_{\alpha/2}]^2 + (N-1)E^2} = \dfrac{200(0.38)(0.62)[1.96]^2}{(0.38)(0.62)[1.96]^2 + (200-1)0.025^2} = 176$

41.   The upper confidence interval limit is greater than 100%. Given that the percentage cannot exceed 100%, change the upper limit to 100%.

$\hat{p} \pm z_{\alpha/2}\sqrt{\dfrac{\hat{p}\hat{q}}{n}} = \dfrac{44}{48} \pm 2.575\sqrt{\dfrac{\left(\frac{44}{48}\right)\left(\frac{4}{48}\right)}{48}}$

$0.814 < p < 1.019$ or $81.4\% < p < 101.9\%$.

43.   The upper confidence interval limit is greater than 100%. Given that the percentage cannot exceed 100%, change the upper limit to 100%.

$\alpha/2 = 0.01/2 = 0.005$. For the upper 0.005, NORM.S.INV(0.995) = 2.576

$$\hat{p} \pm z_{\alpha/2}\sqrt{\frac{\hat{p}\hat{q}}{n}} = \frac{44}{48} \pm 2.576\sqrt{\frac{\left(\frac{44}{48}\right)\left(\frac{4}{48}\right)}{48}} = 0.9167 \pm 0.10276$$

$$0.814 < p < 1.019 \text{ or } 81.4\% < p < 101.9\%.$$

Section 7-3, Basic Skills and Concepts

1. a. $233.4 \text{ sec} < \mu < 256.65 \text{ sec}$

    b.   The best point estimate of μ is $\bar{x} = \dfrac{256.65 + 233.4}{2} = 245.025 \text{ sec}$. The margin of error is

    $E = \dfrac{256.65 - 233.4}{2} = 11.625 \text{ sec}.$

3. We have 95% confidence that the limits of 233.4 sec and 256.65 sec contain the true value of the mean of the population of all duration times.

5. Neither the normal nor the Student t distribution applies.

7. Using Excel's T.INV.2T function to find the critical value, $t_{\alpha/2} =$ T.INV.2T(0.01,39) = 2.7079.

9. Using Excel's CONFIDENCE.T function to find the confidence interval, CONFIDENCE.T(0.02,5.013,50) = 1.7049. $9.808 - 1.7049 < \mu < 9.808 + 1.7049$; $8.103 < \mu < 11.513$

    Because the sample size is greater than 30, the confidence interval yields a reasonable estimate of μ , even though the data appear to be from a population that is not normally distributed.

11. Use Excel's Descriptive Statistics tool to find the confidence interval. Enter the label **Thousands of dollars** in cell A1 followed by the five data values. Click **DATA** at the top of the screen. Select **Descriptive Statistics**. Enter the range of the compensation values including the label (**A1:A6**). Select **Labels in First Row**. Select **Summary statistics**. Select **Confidence Level for Mean: 95%**. Use the mean and confidence level to construct the 95% confidence interval.

Thousands of dollars	
Mean	12898.0002
Standard Error	3452.06191
Median	14765
Mode	#N/A
Standard Deviation	7719.04509
Sample Variance	59583657.1
Kurtosis	2.65655217
Skewness	-1.5622349
Range	19628.999
Minimum	0.001
Maximum	19629
Sum	64490.001
Count	5
Confidence Level(95.0%)	9584.46038

    $12898.00 - 9584.4604 < \mu < 12898.00 + 9584.4604$

    $3{,}313.54 \text{ thousand dollars} < \mu < 22{,}482.46 \text{ thousand dollars}$

    The \$1 salary of Jobs is an outlier that is very far away from the other values, and that outlier has a dramatic effect on the confidence interval.

13. Using Excel's CONFIDENCE.T function to find the confidence interval, CONFIDENCE.T(0.05,0.62,106) = 0.1194.

    $98.20 - 0.1194 < \mu < 98.20 + 0.1194$; $98.08°F < \mu < 98.32°F$

    Because the confidence interval does not contain 98.6°F, it appears that the mean body temperature is not 98.6°F, as is commonly believed.

15. Using Excel's CONFIDENCE.T function to find the confidence interval, CONFIDENCE.T(0.02,21,49) = 7.2197. $0.4 - 7.2197 < \mu < 0.4 + 7.2197$; $-6.8 \text{ mg / dL} < \mu < 7.6 \text{ mg / dL}$

    Because the confidence interval includes the value of 0, it is very possible that the mean of the changes in LDL cholesterol is equal to 0, suggesting that the garlic treatment did not affect LDL cholesterol levels. It does not appear that garlic is effective in reducing LDL cholesterol.

17. Use Excel's Descriptive Statistics tool to find the confidence interval. Enter the label **Millions of dollars** in cell A1 followed by the 14 data values. Click **DATA** at the top of the screen. Select **Descriptive Statistics**. Enter the range of the values including the label (**A1:A17**). Select **Labels in First Row**. Select **Summary statistics**. Select **Confidence Level for Mean: 99%**. Use the mean and confidence level to construct the 99% confidence interval.

$$16.3571 - 11.6904 < \mu < 16.3571 + 11.6904$$

$$4.7 \; million \; dollars < \mu < 28.0 \; million \; dollars$$

The data appear to have a distribution that is far from normal, so the confidence interval might not be a good estimate of the population mean. The population is likely to be the list of box office receipts for each day of the movie's release. Because the values are from the first 14 days of release, the sample values are not a simple random sample, and they are likely to be the largest of all such values, so the confidence interval is not a good estimate of the population mean.

Millions of dollars	
Mean	16.3571
Standard Error	3.88093
Median	10
Mode	10
Standard Deviation	14.5211
Sample Variance	210.863
Kurtosis	4.7359
Skewness	2.02466
Range	54
Minimum	4
Maximum	58
Sum	229
Count	14
Confidence Level(99.0%)	11.6904

19. Use Excel's Descriptive Statistics tool to find the confidence interval. Enter the label **W/kg** in cell A1 followed by the 11 data values. Click **DATA** at the top of the screen. Select **Descriptive Statistics**. Enter the range of the values including the label (**A1:A12**). Select **Labels in First Row**. Select **Summary statistics**. Select **Confidence Level for Mean: 90%**. Use the mean and confidence level to construct the 90% confidence interval.

$$0.9382 - 0.2311 < \mu < 0.9382 + 0.2311$$

$$0.707 \; W / kg < \mu < 1.169 \; W / kg$$

The sample data meet the loose requirement of having a normal distribution. Because the confidence interval is entirely below the standard of 1.6 W/kg, it appears that the mean amount of cell phone radiation is less than the FCC standard, but there could be individual cell phones that exceed the standard.

W/kg	
Mean	0.93818
Standard Error	0.1275
Median	0.92
Mode	#N/A
Standard Deviation	0.42287
Sample Variance	0.17882
Kurtosis	-1.2126
Skewness	0.37762
Range	1.17
Minimum	0.38
Maximum	1.55
Sum	10.32
Count	11
Confidence Level(90.0%)	0.23109

21. Use Excel's Descriptive Statistics tool to find the confidence interval. Enter the label **ug/g** in cell A1 followed by the 10 data values. Click **DATA** at the top of the screen. Select **Descriptive Statistics**. Enter the range of the values including the label (**A1:A11**). Select **Labels in First Row**. Select **Summary statistics**. Select **Confidence Level for Mean: 95%**. Use the mean and confidence level to construct the 95% confidence interval.

$$11.05 - 4.6221 < \mu < 11.05 + 4.6221$$

$$6.43 \; ug / g < \mu < 15.67 \; ug / g$$

The sample data meet the loose requirement of having a normal distribution. We cannot conclude that the population mean is less than 7 μg/g , because the confidence interval shows that the mean might be greater than that level.

ug/g	
Mean	11.05
Standard Error	2.04321
Median	9.5
Mode	20.5
Standard Deviation	6.46121
Sample Variance	41.7472
Kurtosis	-1.3253
Skewness	0.51324
Range	17.5
Minimum	3
Maximum	20.5
Sum	110.5
Count	10
Confidence Level(95.0%)	4.62207

23. For the unsuccessful, $46.9565 - 3.1223 < \mu < 46.9565 + 3.1223$; $43.8 \; years < \mu < 50.1 \; years$

For the successful, $44.5172 - 1.9117 < \mu < 44.5172 + 1.9117$; $42.6 years < \mu < 46.4 \; years$

Although final conclusions about means of populations should not be based on the overlapping of confidence intervals, the confidence intervals do overlap, so it appears that both populations could have the same mean, and there is not clear evidence of discrimination based on age.

25. The sample size is $n = \left[\dfrac{z_{\alpha/2}\sigma}{E}\right]^2 = \left[\dfrac{1.645 \cdot 15}{3}\right]^2 = 68$, and it does appear to be very reasonable.

27. The required sample size is $n = \left[\dfrac{z_{\alpha/2}\sigma}{E}\right]^2 = \left[\dfrac{2.32635 \cdot 2157}{250}\right]^2 = 403$. It is not likely that you would find that many two-year-old used Corvettes in your region.

29. Use $\sigma = \dfrac{2400 - 600}{4} = 450$ to get a sample size of $n = \left[\dfrac{z_{\alpha/2}\sigma}{E}\right]^2 = \left[\dfrac{2.33 \cdot 450}{100}\right]^2 = 110$. The margin of error of 100 points seems too high to provide a good estimate of the mean SAT score.

31. With the range rule of thumb, use $\sigma = \dfrac{90 - 46}{4} = 11$ to get a required sample size of

$n = \left[\dfrac{z_{\alpha/2}\sigma}{E}\right]^2 = \left[\dfrac{1.96 \cdot 11}{2}\right]^2 = 117$.

With $s = 10.3$, the required sample size is $n = \left[\dfrac{z_{\alpha/2}\sigma}{E}\right]^2 = \left[\dfrac{1.96 \cdot 10.3}{2}\right]^2 = 102$. The better estimate of s is the standard deviation of the sample, so the correct sample size is likely to be closer to 102 than 117.

33. Use Excel's Descriptive Statistics tool to find the confidence interval. Click **DATA** at the top of the screen. Select **Descriptive Statistics**. Enter the range of the MAG values including the label. Select **Labels in First Row**. Select **Summary statistics**. Select **Confidence Level for Mean: 99%**. Use the mean and confidence level to construct the 99% confidence interval.

$1.1842 - 0.2226 < \mu < 1.1842 + 0.2226$; $0.962 < \mu < 1.407$

Section 7-3, Beyond the Basics

35. Using Excel's CONFIDENCE.NORM function to find the confidence interval, CONFIDENCE.NORM(0.02,5.013,50) = 1.6493. $9.808 - 1.6493 < \mu < 9.808 + 1.6493$;

$8.159\ km < \mu < 11.457\ km$

37. Using Excel's CONFIDENCE.NORM function to find the confidence interval, CONFIDENCE.NORM(0.05,7718.8,5) = 6765.702. $12898.00 - 6765.702 < \mu < 12898.00 + 6765.702$;

$6132.30\ thousand\ dollars < \mu < 19663.70\ thousand\ dollars$

39. When 3.0 is changed to 300: $40.75 - 65.2951 < \mu < 40.75 + 65.2951$; $-24.55\ ug\,/\,g < \mu < 106.05\ ug\,/\,g$

The sample data do not appear to meet the loose requirement of having a normal distribution. The effect of the outlier on the confidence interval is very substantial. Outliers should be discarded if they are known to be errors. If an outlier is a correct value, it might be very helpful to see its effects by constructing the confidence interval with and without the outlier included.

41.
$x \pm 9.68|3.0|$

$-26.0\ m < \mu < 32.0\ m$

The confidence interval based on the first sample value is much wider than the confidence interval based on all 10 sample values.

Section 7-4, Basic Skills and Concepts

1. $\sqrt{916.591\ (\text{mg/dL})^2} < \sqrt{\sigma^2} < \sqrt{2252.1149\ (\text{mg/dL})^2} \Rightarrow 30.3\ \text{mg/dL} < \sigma < 47.5\ \text{mg/dL}$. We have 95% confidence that the limits of 30.3 mg/dL and 47.5 mg/dL contain the true value of the standard deviation of the LDL cholesterol levels of all women.

3. The original sample values can be identified, but the dotplot shows that the sample appears to be from a population having a uniform distribution, not a normal distribution as required. Because the normality requirement is not satisfied, the confidence interval estimate of s should not be constructed using the methods of this section.

5. $\alpha/2 = 0.01/2 = 0.005$; df $= n - 1 = 25 - 1 = 24$

    Using Excel's CHISQ.INV function, $\chi_L^2 = \text{CHISQ.INV}(0.005,24) = 9.886$; $\chi_R^2 = \text{CHISQ.INV}(0.995,24) = 45.559$.

    $$\sqrt{\frac{(n-1)s^2}{\chi_R^2}} < \sigma < \sqrt{\frac{(n-1)s^2}{\chi_L^2}}$$

    $$\sqrt{\frac{(25-1)0.24^2}{45.559}} < \sigma < \sqrt{\frac{(25-1)0.24^2}{9.886}}$$

    $$0.17\ \text{mg} < \sigma < 0.37\ \text{mg}$$

7. $\alpha/2 = 0.05/2 = 0.025$; df $= n - 1 = 40 - 1 = 39$

    Using Excel's CHISQ.INV function, $\chi_L^2 = \text{CHISQ.INV}(0.025,39) = 23.654$; $\chi_R^2 = \text{CHISQ.INV}(0.975,39) = 58.120$.

    $$\sqrt{\frac{(n-1)s^2}{\chi_R^2}} < \sigma < \sqrt{\frac{(n-1)s^2}{\chi_L^2}}$$

    $$\sqrt{\frac{(40-1)65.2^2}{58.120}} < \sigma < \sqrt{\frac{(40-1)65.2^2}{23.654}}$$

    $$53.409 < \sigma < 83.720$$

9. Using Excel's CHISQ.INV function, $\chi_L^2 = \text{CHISQ.INV}(0.05,105) = 82.354$; $\chi_R^2 = \text{CHISQ.INV}(0.95,105) = 129.918$.

    $$\sqrt{\frac{(n-1)s^2}{\chi_R^2}} < \sigma < \sqrt{\frac{(n-1)s^2}{\chi_L^2}}$$

    $$\sqrt{\frac{(106-1)0.62^2}{129.918}} < \sigma < \sqrt{\frac{(106-1)0.62^2}{82.354}}$$

    $$0.557°\text{F} < \sigma < 0.700°\text{F}$$

11. Using Excel's CHISQ.INV function, $\chi_L^2 = \text{CHISQ.INV}(0.005,23) = 9.260$; $\chi_R^2 = \text{CHISQ.INV}(0.995,23) = 44.181$.

    $$\sqrt{\frac{(n-1)s^2}{\chi_R^2}} < \sigma < \sqrt{\frac{(n-1)s^2}{\chi_L^2}}$$

    $$\sqrt{\frac{(24-1)42.8^2}{44.181}} < \sigma < \sqrt{\frac{(24-1)42.8^2}{9.260}}$$

    $$30.9\ \text{mL} < \sigma < 67.45\ \text{mL}$$

The confidence interval shows that the standard deviation is not likely to be less than 30 mL, so the variation is too high instead of being at an acceptable level below 30 mL. (Such one-sided claims should be tested using the formal methods presented in Chapter 8.)

13. $s = 0.36576$; $\chi_L^2 = $ CHISQ.INV(0.05,6) = 1.635; $\chi_R^2 = $ CHISQ.INV(0.95,6) = 12.592.

$$\sqrt{\frac{(n-1)s^2}{\chi_R^2}} < \sigma < \sqrt{\frac{(n-1)s^2}{\chi_L^2}}$$

$$\sqrt{\frac{(7-1)0.36576^2}{12.592}} < \sigma < \sqrt{\frac{(7-1)0.36576^2}{1.635}}$$

$$0.252 \text{ ppm} < \sigma < 0.701 \text{ ppm}$$

15. $s = 7.22$ and df $= n - 1 = 25 - 1 = 24$; $\chi_L^2 = $ CHISQ.INV(0.005,24) = 9.886; $\chi_R^2 = $ CHISQ.INV(0.995,24) = 45.559.

$$\sqrt{\frac{(n-1)s^2}{\chi_R^2}} < \sigma < \sqrt{\frac{(n-1)s^2}{\chi_L^2}}$$

$$\sqrt{\frac{(25-1)7.22^2}{45.559}} < \sigma < \sqrt{\frac{(25-1)7.22^2}{9.886}}$$

$$5.2 \text{ years} < \sigma < 11.5 \text{ years}$$

Successful applicants: $s = 5.026$ and df $= n - 1 = 29 - 1 = 28$; $\chi_L^2 = $ CHISQ.INV(0.005,28) = 12.461; $\chi_R^2 = $ CHISQ.INV(0.995,28) = 50.993.

$$\sqrt{\frac{(n-1)s^2}{\chi_R^2}} < \sigma < \sqrt{\frac{(n-1)s^2}{\chi_L^2}}$$

$$\sqrt{\frac{(29-1)5.026^2}{50.993}} < \sigma < \sqrt{\frac{(29-1)5.026^2}{12.461}}$$

$$3.7 \text{ years} < \sigma < 7.5 \text{ years}$$

Although final conclusions about means of populations should not be based on the overlapping of confidence intervals, the confidence intervals do overlap, so it appears that the two populations have standard deviations that are not dramatically different.

17. $s = 0.0165$ and df $= n - 1 = 37 - 1 = 36$; $\chi_L^2 = $ CHISQ.INV(0.1,36) = 19.233; $\chi_R^2 = $ CHISQ.INV(0.99,36) = 58.619.

$$\sqrt{\frac{(n-1)s^2}{\chi_R^2}} < \sigma < \sqrt{\frac{(n-1)s^2}{\chi_L^2}}$$

$$\sqrt{\frac{(37-1)0.0165^2}{58.619}} < \sigma < \sqrt{\frac{(37-1)0.0165^2}{19.233}}$$

$$0.01293 \text{ g} < \sigma < 0.02257 \text{ g}$$

19. 33,218 is too large. There aren't 33,218 statistics professors in the population, and even if there were, that sample size is too large to be practical.

21. The sample size is 768. Because the population does not have a normal distribution, the computed minimum sample size is not likely to be correct.

Section 7-4, Beyond the Basics

23. $\chi_L^2 = \frac{1}{2}\left[-z_{\alpha/2} + \sqrt{2k-1}\right]^2 = \frac{1}{2}\left[-1.645 + \sqrt{2\cdot 105-1}\right]^2 = 82.072$ and

$\chi_R^2 = \frac{1}{2}\left[z_{\alpha/2} + \sqrt{2k-1}\right]^2 = \frac{1}{2}\left[1.645 + \sqrt{2\cdot 105-1}\right]^2 = 129.635$

(Excel using $z_{\alpha/2} = 1.644853626$: $\chi_L^2 = 82.073$ and $\chi_R^2 = 129.632$). The approximate values are quite close to the actual critical values.

Chapter Quick Quiz

1. $40\% - 3.1\% < p < 40\% + 3.1\%$

   $36.9\% < p < 43.1\%$

2. $\hat{p} = \dfrac{0.511 + 0.449}{2} = 0.480$

3. We have 95% confidence that the limits of 0.449 and 0.511 contain the true value of the proportion of females in the population of medical school students.

4. $z = 1.645$

5. $n = \dfrac{\left[z_{\alpha/2}\right]^2 \hat{p}\hat{q}}{E^2} = \dfrac{[1.645]^2 (0.25)}{0.03^2} = 752$

6. $n = \left[\dfrac{z_{\alpha/2}\sigma}{E}\right]^2 = \left[\dfrac{2.575 \cdot 15}{2}\right]^2 = 373$ (Tech: 374)

7. The sample must be a simple random sample and there is a loose requirement that the sample values appear to be from a normally distributed population.

8. The degrees of freedom is the number of sample values that can vary after restrictions have been imposed on all of the values. For the sample data in Exercise 7, df = 5.

9. Using Excel's T.INV.2T function, T.INV.2T(0.05,5) = 2.5706.

10. Using Excel's CHISQ.INV function, χ_L^2 = CHISQ.INV(0.025,5) = 0.831; χ_R^2 = CHISQ.INV(0.975,5) = 12.833.

Review Exercises

1. a. $\hat{p} = \dfrac{284}{557} = 0.510 = 51.0\%$

   b. $\hat{p} \pm z_{\alpha/2}\sqrt{\dfrac{\hat{p}\hat{q}}{n}} = \dfrac{284}{557} \pm 1.96\sqrt{\dfrac{\left(\frac{284}{557}\right)\left(\frac{273}{557}\right)}{557}}$

      $46.8\% < p < 55.1\%$

   c. No, the confidence interval shows that the population percentage might be 50% or less, so we cannot safely conclude that the majority of adults say that they are underpaid.

2. $n = \dfrac{\left[z_{\alpha/2}\right]^2 \hat{p}\hat{q}}{E^2} = \dfrac{[2.575829]^2 (0.25)}{0.02^2} = 4147$

3. $n = \left[\dfrac{z_{\alpha/2}\sigma}{E}\right]^2 = \left[\dfrac{2.326348 \cdot 16}{3}\right]^2 = 154$

4.   a.   Student t distribution

b.   Normal distribution

c.   The distribution is not normal, Student t, or chi-square.

d. χ^2 (chi-square distribution)

e.   Normal distribution

5.   a. $\alpha/2 = 0.02/2 = 0.01$. For the upper 0.10, NORM.S.INV(0.99) = 2.326348.

$$n = \dfrac{\left[z_{\alpha/2}\right]^2 \hat{p}\hat{q}}{E^2} = \dfrac{\left[2.326348\right]^2 (0.25)}{0.05^2} = 542$$

b. $n = \left[\dfrac{z_{\alpha/2}\sigma}{E}\right]^2 = \left[\dfrac{2.326348 \cdot 337}{50}\right]^2 = 246$

c.   542

6. $\alpha/2 = 0.10/2 = 0.05$. For the upper 0.05, NORM.S.INV(0.95) = 1.644854.

$$\hat{p} \pm z_{\alpha/2}\sqrt{\dfrac{\hat{p}\hat{q}}{n}} = 0.64 \pm 1.644854\sqrt{\dfrac{(0.64)(0.36)}{1011}}$$
$$61.5\% < p < 66.5\%$$

Because the entire confidence interval is above 50%, we can safely conclude that the majority of adults consume alcoholic beverages.

7.   Enter the data in an Excel worksheet. Then use Excel's Descriptive Statistics tool to find the confidence interval. Click **DATA** at the top of the screen. Select **Descriptive Statistics**. Enter the range of the discrepancy values. Select **Labels in First Row** if you included a label in the first row. Select **Summary statistics**. Select **Confidence Level for Mean: 95%**. Use the mean and confidence level to construct the 95% confidence interval. Mean = 143 and Confidence Level = 165.0534.

$143 - 165.0534 < \mu < 143 + 165.0534$; $-22.05 \sec < \mu < 308.05 \sec$

8.   Using Excel's T.INV.2T function, T.INV.2T(0.10,39) = 1.684875.

$$\bar{x} \pm t_{\alpha/2}\dfrac{s}{\sqrt{n}} = 7.15 \pm 1.684875 \cdot \dfrac{2.28}{\sqrt{40}}$$
$$6.54 < \mu < 7.76$$

Because women and men have some notable physiological differences, the confidence interval does not necessarily serve as an estimate of the mean white blood cell count of men.

9.   Using Excel's T.INV.2T function, T.INV.2T(0.05,6) = 2.446912.

$$\bar{x} \pm t_{\alpha/2}\dfrac{s}{\sqrt{n}} = 42.7 \pm 2.446912 \cdot \dfrac{5.6}{\sqrt{7}}$$
$$37.5 \text{ g} < \mu < 47.9 \text{ g}$$

10. $s = 5.6$ and df $= n - 1 = 7 - 1 = 6$; $\chi_L^2 = $ CHISQ.INV(0.025,6) = 1.237; $\chi_R^2 = $ CHISQ.INV(0.975,6) = 14.449.

$$\sqrt{\frac{(n-1)s^2}{\chi_R^2}} < \sigma < \sqrt{\frac{(n-1)s^2}{\chi_L^2}}$$

$$\sqrt{\frac{(7-1)5.6^2}{14.449}} < \sigma < \sqrt{\frac{(7-1)5.6^2}{1.237}}$$

$$3.6 \text{ g} < \sigma < 12.3 \text{ g}$$

Cumulative Review Exercises

1. Use Excel's Descriptive Statistics tool to find the confidence interval. Enter the label **Frequency** in cell A1 followed by the six data values. Click **DATA** at the top of the screen. Select **Descriptive Statistics**. Enter the range of the frequencies including the label (**A1:A7**). Select **Labels in First Row**. Select **Summary statistics**. A portion of the output is displayed at the right. Mean = 5.5, Median = 5, and Standard Deviation = 3.7815.

Frequency	
Mean	5.5
Median	5
Standard Deviation	3.78153
Count	6

2. The range of usual values is from $5.5 - 2(3.8) = -2.1$ to $5.5 + 2(3.8) = 3.8$ (or from 0 to 13.1).

3. Ratio level of measurement; discrete data.

4. $n = \left[\frac{z_{\alpha/2}\sigma}{E} \right]^2 = \left[\frac{1.96 \cdot 5.8}{2} \right]^2 = 33$ campuses

5. Using Excel's T.INV.2T function, T.INV.2T(0.05,39) = 2.02269.

$$\bar{x} \pm t_{\alpha/2} \frac{s}{\sqrt{n}} = 5.5 \pm 2.02269 \cdot \frac{5.8}{\sqrt{40}}$$

$$3.6 < \mu < 7.4$$

The population should include only colleges of the same type as the sample, so the population consists of all large urban campuses with residence halls.

6. The graphs suggest that the population has a distribution that is skewed (to the right) instead of being normal. The histogram shows that some taxi-out times can be very long, and that can occur with heavy traffic, but little or no traffic cannot make the taxi-out time very low. There is a minimum time required, regardless of traffic conditions. Construction of a confidence interval estimate of a population standard deviation has a strict requirement that the sample data are from a normally distributed population, and the graphs show that this strict normality requirement is not satisfied.

7. a. $\hat{p} \pm z_{\alpha/2} \sqrt{\frac{\hat{p}\hat{q}}{n}} = 0.59 \pm 1.96 \sqrt{\frac{(0.59)(0.31)}{1003}}$ (or $0.560 < p < 0.621$ if using $x = 592$)

$$0.560 < p < 0.620$$

b. Because the survey was about shaking hands and because it was sponsored by a supplier of hand sanitizer products, the sponsor could potentially benefit from the results, so there might be some pressure to obtain results favorable to the sponsor.

c. $n = \frac{[z_{\alpha/2}]^2 \hat{p}\hat{q}}{E^2} = \frac{[1.96]^2(0.25)}{0.025^2} = 1083$

8. There does not appear to be a correlation between HDL and LDL cholesterol levels.

9. a. Using Excel's NORM.DIST function, $1 - \text{NORM.DIST}(185,175,9,\text{TRUE}) = 1 - 0.8667 = 13.33\%$

b. 5th percentile using Excel's NORM.INV function, NORM.INV(0.05,175,9) = 160.2 mm.

95th percentile using Excel's NORM.INV function, NORM.INV(0.95,175,9) = 189.8 mm.

10. a.   There are 10^3 possible tickets so the probability of winning by purchasing one ticket is $\dfrac{1}{1000}$.

   b. $1 - \dfrac{1}{1000} = \dfrac{999}{1000}$.

   c. $\left(\dfrac{999}{1000}\right)^{10} = 0.990$.

Chapter 8

Hypothesis Testing

Section 8-2, Basic Skills and Concepts

1. Rejection of the aspirin claim is more serious because the aspirin is a drug treatment. The wrong aspirin dosage can cause adverse reactions. M&Ms do not have those same adverse reactions. It would be wise to use a smaller significance level for testing the aspirin claim.

3. a. $H_0: \mu = 98.6°F$
                                                        b. $H_1: \mu \neq 98.6°F$

   c. Reject the null hypothesis or fail to reject the null hypothesis.

   d. No. In this case, the original claim becomes the null hypothesis. For the claim that the mean body temperature is equal to 98.6°F, we can either reject that claim or fail to reject it, but we cannot state that there is sufficient evidence to support that claim.

5. a. $p = 0.20$
                                                   b. $H_0: p = 0.20$ and $H_1: p \neq 0.20$

7. a. $\mu \leq 76$
                                                   b. $H_0: \mu = 76$ and $H_1: \mu < 76$

9. There is not sufficient evidence to warrant rejection of the claim that 20% of adults smoke.

11. There is not sufficient evidence to warrant rejection of the claim that the mean pulse rate of adult females is 76 or lower.

13. $z = \dfrac{\hat{p} - p}{\sqrt{\frac{pq}{n}}} = \dfrac{0.89 - 0.75}{\sqrt{\frac{(0.75)(0.25)}{1021}}} = 10.33$ (or $z = 10.35$ if using $x = 909$)

15. $\chi^2 = \dfrac{(n-1)s^2}{\sigma^2} = \dfrac{(40-1)2.28^2}{5^2} = 8.110$

In Exercises 17-23, you will be finding *P*-values and critical values. You will use Excel's NORM.S.DIST function to find the *P*-value and Excel's NORM.S.INV function to find the critical value. Recall that, in Excel, formulas begin with an equal sign.

17. *P*-Value: **=1 – NORM.S.DIST(2.00,TRUE)** = 0.0228. Critical value: **=NORM.S.INV(0.95)** = 1.645.

19. *P*-Value: **=2*NORM.S.DIST(-1.75,TRUE)** = 0.0801. Critical values: **=NORM.S.INV(0.025)** = -1.96 and **=NORM.S.INV(0.975)** = 1.96.

21. *P*-Value: **=2*NORM.S.DIST(-1.23, TRUE)** = 0.2187. Critical values: **=NORM.S.INV(0.025)** = -1.96 and **=NORM.S.INV(0.975)** = 1.96.

23. *P*-Value: **=NORM.S.DIST(-3.00, TRUE)** = 0.00135. Critical value: **=NORM.S.INV(0.05)** = -1.645.

25. a. Reject H_0.

   b. There is sufficient evidence to support the claim that the percentage of blue M&Ms is greater than 5%.

27. a. Fail to reject H_0.

   b. There is not sufficient evidence to warrant rejection of the claim that women have heights with a mean equal to 160.00 cm.

29. a. $H_0: p = 0.5$ and $H_1: p > 0.5$
                                                   b. $\alpha = 0.01$

   c. Normal distribution.
                                                   d. Right-tailed.

   e. $z = 1.00$

   f. Using Excel's NORM.S.DIST function, the *P*-value **=1 – NORM.S.DIST(1.00,TRUE)** = 0.1587.

   g. Using Excel's NORM.S.INV function, the critical value **=NORM.S.INV(0.99)** = 2.33.

   h. 0.01

31. Type I error: In reality $p = 0.1$, but we reject the claim that $p = 0.1$. Type II error: In reality $p \neq 0.1$, but we fail to reject the claim that $p = 0.1$.

33. Type I error: In reality $p = 0.5$, but we support the claim that $p > 0.5$. Type II error: In reality $p > 0.5$, but we fail to support that conclusion.

Section 8-2, Beyond the Basics

35. The power of 0.96 shows that there is a 96% chance of rejecting the null hypothesis of $p = 0.08$ when the true proportion is actually 0.18. That is, if the proportion of Chantix users who experience abdominal pain is actually 0.18, then there is a 96% chance of supporting the claim that the proportion of Chantix users who experience abdominal pain is greater than 0.08.

37. From $p = 0.5$, $\hat{p} = 0.5 + 1.645\sqrt{\dfrac{(0.5)(0.5)}{n}}$, from $p = 0.55$, $\hat{p} = 0.55 - 0.842\sqrt{\dfrac{(0.55)(0.45)}{n}}$; Since

$[P(z > -0.842) = 0.8000]$, so:

$$0.5 + 1.645\sqrt{\frac{(0.5)(0.5)}{n}} = 0.55 - 0.842\sqrt{\frac{(0.55)(0.45)}{n}}$$

$$0.5\sqrt{n} + 1.645\sqrt{0.25} = 0.55\sqrt{n} - 0.842\sqrt{0.2475}$$

$$0.05\sqrt{n} = 1.645\sqrt{0.25} + 0.842\sqrt{0.2475}$$

$$n = \left(\frac{1.645\sqrt{0.25} + 0.842\sqrt{0.2475}}{0.05}\right)^2 = 617$$

Section 8-3, Basic Skills and Concepts

1. The P-value method and the critical value method always yield the same conclusion. The confidence interval method might or might not yield the same conclusion obtained by using the other two methods.

3. P-value = 0.00000000550. Because the P-value is so low, we have sufficient evidence to support the claim that $p < 0.5$.

5.   a.   Left-tailed   b. $z = -1.94$
     c. P-value = 0.0260 (rounded)   d. H_0: $p = 0.1$. Reject the null hypothesis.
     e.   There is sufficient evidence to support the claim that less than 10% of treated subjects experience headaches.

7.   a.   Two-tailed   b. $z = -0.82$
     c. P-value = 0.4106   d. H_0: $p = 0.35$. Fail to reject the null hypothesis.
     e.   There is not sufficient evidence to warrant rejection of the claim that 35% of adults have heard of the Sony Reader.

9. H_0: $p = 0.25$. H_1: $p \neq 0.25$. Test statistic: $z = \dfrac{\frac{152}{580} - 0.25}{\sqrt{\frac{(0.25)(0.75)}{580}}} = 0.67$. Using Excel's NORM.S.INV function, the critical values are =**NORM.S.INV(0.005)** = -2.576 and =**NORM.S.INV(0.995)** = 2.576. Using Excel's NORM.S.DIST function, the P-value =**2*NORM.S.DIST(-0.67, TRUE)** = 0.5029. Fail to reject H_0. There is not sufficient evidence to warrant rejection of the claim that 25% of offspring peas will be yellow.

11. H_0: $p = 0.5$. H_1: $p > 0.5$. Test statistic: $z = \dfrac{\frac{531}{1002} - 0.5}{\sqrt{\frac{(0.5)(0.5)}{1002}}} = 1.90$. Using Excel's NORM.S.INV function, the

critical value is =**NORM.S.INV(0.95)** = 1.645. Using Excel's NORM.S.DIST function, the P-value =**NORM.S.DIST(-1.90, TRUE)** = 0.0287. Reject H_0. There is sufficient evidence to support the claim that the majority of adults feel vulnerable to identify theft.

13. H_0: $p = 0.5$. H_1: $p > 0.5$. Test statistic: $z = \dfrac{\frac{879}{945} - 0.5}{\sqrt{\frac{(0.5)(0.5)}{945}}} = 26.45$. Using Excel's NORM.S.INV function, the

critical value is =**NORM.S.INV(0.99)** = 2.326. Using Excel's NORM.S.DIST function, the P-value =**NORM.S.DIST(-26.45, TRUE)** = 1.8233E-154. Reject H_0. There is sufficient evidence to support the claim that the XSORT method is effective in increasing the likelihood that a baby will be a girl.

15. H_0: $p = 0.5$. H_1: $p \neq 0.5$. Test statistic: $z = \dfrac{\frac{123}{280} - 0.5}{\sqrt{\frac{(0.5)(0.5)}{280}}} = -2.03$. Using Excel's NORM.S.INV function, the

critical values are =**NORM.S.INV(0.05)** = -1.645 and =**NORM.S.INV(0.95)** = 1.645. Using Excel's NORM.S.DIST function, the P-value =**2*NORM.S.DIST(-2.03, TRUE)** = 0.0424. Reject H_0. There is sufficient evidence to warrant rejection of the claim that touch therapists use a method equivalent to random guesses. However, their success rate of 123/280 (or 43.9%) indicates that they performed worse than random guesses, so they do not appear to be effective.

17. H_0: $p = \frac{1}{3}$. H_1: $p < \frac{1}{3}$. Test statistic: $z = \dfrac{\frac{172}{611} - \frac{1}{3}}{\sqrt{\frac{(\frac{1}{3})(\frac{2}{3})}{611}}} = -2.72$. Using Excel's NORM.S.INV function, the critical

value is =**NORM.S.INV(0.01)** = -2.326. Using Excel's NORM.S.DIST function, the P-value =**NORM.S.DIST(-2.72, TRUE)** = 0.0033. Reject H_0. There is sufficient evidence to support the claim that fewer than 1/3 of the challenges are successful. Players don't appear to be very good at recognizing referee errors.

19. H_0: $p = 0.000340$. H_1: $p \neq 0.000340$. Test statistic: $z = \dfrac{\frac{135}{420,095} - 0.000340}{\sqrt{\frac{(0.000340)(0.99966)}{420,095}}} = -0.66$. Using Excel's

NORM.S.INV function, the critical values are =**NORM.S.INV(0.0025)** = -2.807 and =**NORM.S.INV(0.9975)** = 2.807. Using Excel's NORM.S.DIST function, the P-value =**2*NORM.S.DIST(-0.66, TRUE)** = 0.5093. Fail to reject H_0. There is not sufficient evidence to support the claim that the rate is different from 0.0340%. Cell phone users should not be concerned about cancer of the brain or nervous system.

21. H_0: $p = 0.5$. H_1: $p \neq 0.5$. Test statistic: $z = \dfrac{\frac{235}{414} - 0.5}{\sqrt{\frac{(0.5)(0.5)}{414}}} = 2.75$. Using Excel's NORM.S.INV function, the

critical values are =**NORM.S.INV(0.025)** = -1.96 and =**NORM.S.INV(0.975)** = 1.96. Using Excel's NORM.S.DIST function, the P-value =**2*NORM.S.DIST(-2.75, TRUE)** = 0.0060. Reject H_0. There is sufficient evidence to warrant rejection of the claim that the coin toss is fair in the sense that neither team has an advantage by winning it. The coin toss rule does not appear to be fair.

23. H_0: $p = 0.5$. H_1: $p < 0.5$. Test statistic: $z = \dfrac{\frac{152}{380} - 0.5}{\sqrt{\frac{(0.5)(0.5)}{380}}} = -3.90$. Using Excel's NORM.S.INV function, the

critical value is =**NORM.S.INV(0.01)** = -2.326. Using Excel's NORM.S.DIST function, the P-value =**NORM.S.DIST(-3.90, TRUE)** = 4.80963E-05. Reject H_0. There is sufficient evidence to support the claim that fewer than half of smartphone users identify the smartphone as the only thing they could not live without. Because only smartphone users were surveyed, the results do not apply to the general population.

25. H_0: $p = 0.25$. H_1: $p > 0.25$. Test statistic: $z = \dfrac{0.29 - 0.25}{\sqrt{\frac{(0.25)(0.75)}{427}}} = 1.91$ or $z = 1.93$ (using x = 124). Using Excel's

NORM.S.INV function, the critical value is =**NORM.S.INV(0.95)** = 1.645 (assuming a 0.05 significance level). Using Excel's NORM.S.DIST function, the P-value =**NORM.S.DIST(-1.91, TRUE)** = 0.0281 using 1.91 or the P-value =**NORM.S.DIST(-1.93, TRUE)** = 0.0269 using 1.93. Reject H_0. There is sufficient evidence to support the claim that more than 25% of women purchase books online.

27. H_0: $p = 0.75$. H_1: $p > 0.75$. Test statistic: $z = \dfrac{0.90 - 0.75}{\sqrt{\frac{(0.75)(0.25)}{514}}} = 7.85$ or $z = 7.89$ (using x = 463). Using Excel's

NORM.S.INV function, the critical value is =**NORM.S.INV(0.99)** = 2.326. Using Excel's NORM.S.DIST function, the P-value =**NORM.S.DIST(-7.85, TRUE)** = 2.08019E-15 using 7.85 or the P-value =**NORM.S.DIST(-7.89, TRUE)** = 1.51093E-15 using 7.89. Reject H_0. There is sufficient evidence to support the claim that more than 3/4 of all human resource professionals say that the appearance of a job applicant is most important for a good first impression.

29. H_0: $p = 0.791$. H_1: $p < 0.791$. Test statistic: $z = \dfrac{0.39 - 0.791}{\sqrt{\frac{(0.791)(0.209)}{870}}} = -29.09$ or $z = -29.11$ (using x = 339). Using

Excel's NORM.S.INV function, the critical value is =**NORM.S.INV(0.01)** = -2.326. Using Excel's NORM.S.DIST function, the P-value =**NORM.S.DIST(-29.09, TRUE)** = 2.4019E-186 using -29.09 or the P-value =**NORM.S.DIST(-29.11, TRUE)** = 1.3412E-186 using -29.11. Reject H_0. There is sufficient evidence to support the claim that the percentage of selected Americans of Mexican ancestry is less than 79.1%, so the jury selection process appears to be unfair.

31. H_0: $p = 0.75$. H_1: $p > 0.75$. Test statistic: $z = \dfrac{0.77 - 0.75}{\sqrt{\frac{(0.75)(0.25)}{25,000}}} = 7.30$. Using Excel's NORM.S.INV function, the

critical value is =**NORM.S.INV(0.99)** = 2.326. Using Excel's NORM.S.DIST function, the P-value =**NORM.S.DIST(-7.30, TRUE)** = 1.43884E-13. Reject H_0. There is sufficient evidence to support the claim that more than 75% of television sets in use were tuned to the Super Bowl.

33. Among 100 M&Ms, 19 are green. H_0: $p = 0.16$. H_1: $p \neq 0.16$. Test statistic: $z = \dfrac{0.19 - 0.16}{\sqrt{\frac{(0.16)(0.84)}{100}}} = 0.82$. Using

Excel's NORM.S.INV function, the critical values are =**NORM.S.INV(0.025)** = -1.96 and =**NORM.S.INV(0.975)** = 1.96. Using Excel's NORM.S.DIST function, the P-value =**2*NORM.S.DIST(-0.82, TRUE)** = 0.4122. Fail to reject H_0. There is not sufficient evidence to warrant rejection of the claim that 16% of plain M&M candies are green.

Section 8-3, Beyond the Basics

35. H_0: $p = 0.5$. H_1: $p > 0.5$. Using Excel's BINOM.DIST function, the P-value =BINOM.DIST(7,8,.5,FALSE) = 0.03125. Reject H_0. There is sufficient evidence to support the claim that the coin favors heads.

37. a. From $p = 0.40$, $\hat{p} = 0.4 - 1.645\sqrt{\dfrac{(0.4)(0.6)}{50}} = 0.286$

    From $p = 0.25$, $z = \dfrac{0.286 - 0.25}{\sqrt{\frac{(0.25)(0.75)}{50}}} = 0.588$; Power $= P(z < 0.588)$. Using Excel's NORM.S.DIST

    function, power =NORM.S.DIST(0.588,TRUE) = 0.7217.

    b. $1 - 0.7217 = 0.2783$

c.   The power of 0.7217 shows that there is a reasonably good chance of making the correct decision of rejecting the false null hypothesis. It would be better if the power were even higher, such as greater than 0.8 or 0.9.

Section 8-4, Basic Skills and Concepts

1.   The requirements are (1) the sample must be a simple random sample, and (2) either or both of these conditions must be satisfied: The population is normally distributed or $n > 30$. There is not enough information given to determine whether the sample is a simple random sample. Because the sample size is not greater than 30, we must check for normality, but the value of 583 sec appears to be an outlier, and a normal quantile plot or histogram suggests that the sample does not appear to be from a normally distributed population.

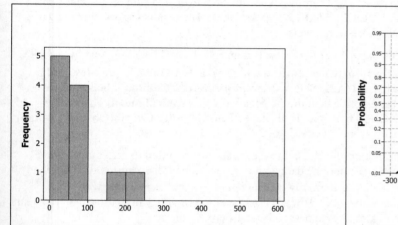 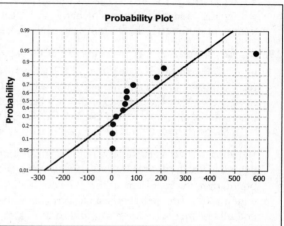

3.   A t test is a hypothesis test that uses the Student t distribution, such as the method of testing a claim about a population mean as presented in this section. The t test methods are much more likely to be used than the z test methods because the t test does not require a known value of σ, and realistic hypothesis tests of claims about μ typically involve a population with an unknown value of σ.

5.   Using Excel's T.DIST.RT function, the P-value =**T.DIST.RT(3.349,24** = 0.0013.

7.   Using Excel's T.DIST.2T function, the P-value =**T.DIST.2T(2.242,20)** = 0.0365.

9. H_0: $\mu = 24$. H_1: $\mu < 24$. Test statistic: $t = -7.323$. Using Excel's T.INV function, the critical value =T.INV(0.05,39) = -1.685. Using Excel's T.DIST function, the P-value =**T.DIST(-7.323,39,TRUE)** = 3.87531E-09. Reject H_0. There is sufficient evidence to support the claim that Chips Ahoy reduced-fat cookies have a mean number of chocolate chips that is less than 24 (but this does not provide conclusive evidence of reduced fat).

11. H_0: $\mu = 33$ years. H_1: $\mu \neq 33$ years. Test statistic: $t = \dfrac{35.9 - 33}{11.1/\sqrt{82}} = 2.367$. Using Excel's T.INV.2T function, the critical values =**T.INV.2T(0.01,81)** = ±2.638. Using Excel's T.DIST.2T function, the P-value =**T.DIST.2T(2.367,81)** = 0.0203. Fail to reject H_0. There is not sufficient evidence to warrant rejection of the claim that the mean age of actresses when they win Oscars is 33 years.

13. H_0: $\mu = 0.8535$ g. H_1: $\mu \neq 0.8535$ g. Test statistic: $t = \dfrac{0.8635 - 0.8535}{0.0570/\sqrt{19}} = 0.765$. Using Excel's T.INV.2T function, the critical value =**T.INV.2T(0.05,18)** = ±2.101. Using Excel's T.DIST.2T function, the P-value =**T.DIST.2T(0.765,18)** = 0.4542. Fail to reject H_0. There is not sufficient evidence to warrant rejection of the

claim that the mean weight of all green M&Ms is equal to 0.8535 g. The green M&Ms do appear to have weights consistent with the package label.

15. H_0: $\mu = 0$ lb. H_1: $\mu > 0$ lb. Test statistic: $t = \dfrac{3 - 0}{4.9/\sqrt{40}} = 3.872$. Using Excel's T.INV function, the critical

value **=T.INV(0.01,39)** = 2.426. Using Excel's T.DIST.RT function, the P-value **=T.DIST.RT(3.872,39)** = 0.0002. Reject H_0. There is sufficient evidence to support the claim that the mean weight loss is greater than 0. Although the diet appears to have statistical significance, it does not appear to have practical significance, because the mean weight loss of only 3.0 lb. does not seem to be worth the effort and cost.

17. H_0: $\mu = 0$. H_1: $\mu > 0$. Test statistic: $t = \dfrac{0.4 - 0}{21.0/\sqrt{49}} = 0.133$. Using Excel's T.INV function, the critical value

=T.INV(0.95,48) = 1.677 (assuming a 0.05 significance level). Using Excel's T.DIST.RT function, the P-value **=T.DIST.RT(0.133,48)** = 0.4474. Fail to reject H_0. There is not sufficient evidence to support the claim that with garlic treatment, the mean change in LDL cholesterol is greater than 0. The results suggest that the garlic treatment is not effective in reducing LDL cholesterol levels.

19. H_0: $\mu = 4$ years. H_1: $\mu > 4$ years. Enter the label **Years** in cell A1 of an Excel worksheet followed by the data given in Exercise 19. Select the **XLSTAT** add-in. Select **Parametric tests**. Select **One-sample *t*-test and *z*-test**. Enter the data range including the label, **A1:A21**. Select **One sample**, **Sheet**, **Column labels**, and **Student's *t*-test**. Click the **Options** tab. Alternative hypothesis: **Mean 1 > Theoretical mean**. Theoretical mean: **4**. Significance level (%): **1**. Click **OK**. The output is displayed below. Test statistic [*t* (Observed value)] = 3.1892. *t* (Critical value) = 2.5395. *P*-value (one-tailed) = 0.0024.

Reject H_0. There is sufficient evidence to support the claim that the mean time required to earn a bachelor's degree is greater than 4.0 years. Because $n \leq 30$ and the data do not appear to be from a normally distributed population, the requirement that "the population is normally distributed or $n > 30$" is not satisfied, so the conclusion from the hypothesis test might not be valid. However, some of the sample values are equal to 4 years and others are greater than 4 years, so the claim does appear to be justified.

One-sample t-test / Upper-tailed test:	
99% confidence interval on the mean:	
(4.5093, +Inf)	
Difference	2.5000
t (Observed value)	3.1892
t (Critical value)	2.5395
DF	19
p-value (one-tailed)	0.0024
alpha	0.01

21. The sample data meet the loose requirement of having a normal distribution. H_0: $\mu = 14$ mg/g.

H_1: $\mu < 14$ mg/g. Enter the label **mg/g** in cell A1 of an Excel worksheet followed by the data given in Exercise 21. Select the **XLSTAT** add-in. Select **Parametric tests**. Select **One-sample *t*-test and *z*-test**. Enter the data range including the label, **A1:A11**. Select **One sample**, **Sheet**, **Column labels**, and **Student's *t*-test**. Click the **Options** tab. Alternative hypothesis: **Mean 1 < Theoretical mean**. Theoretical mean: **14**. Significance level (%): **5**. Click **OK**. The output is displayed on the next page. Test statistic [t (Observed value)] = −1.4438. *t* (Critical value) = −1.8331. *P*-value (one-tailed) = 0.0913.

Fail to reject H_0. There is not sufficient evidence to support the claim that the mean lead concentration for all such medicines is less than 14 mg/g.

One-sample t-test / Lower-tailed test:	
95% confidence interval on the mean:	
(-Inf , 14.7954)	
Difference	-2.9500
t (Observed value)	-1.4438
t (Critical value)	-1.8331
DF	9
p-value (one-tailed)	0.0913
alpha	0.05

23. The sample data meet the loose requirement of having a normal distribution. H_0: $\mu = 63.8$ in.

H_1: $\mu > 63.8$ in. Enter the label **Inches** in cell A1 of an Excel worksheet followed by the data given in Exercise 23. Select the **XLSTAT** add-in. Select **Parametric tests**. Select **One-sample t-test and z-test**. Enter the data range including the label, **A1:A11**. Select **One sample, Sheet, Column labels**, and **Student's t-test**. Click the **Options** tab. Alternative hypothesis: **Mean 1 > Theoretical mean**. Theoretical mean: **63.8**. Significance level (%): **1**. Click **OK**. The output is displayed below. Test statistic [t (Observed value)] = 23.8237. t (Critical value) = 2.8234. *P*-value (one-tailed) <0.0001.

Reject H_0. There is sufficient evidence to support the claim that supermodels have heights with a mean that is greater than the mean height of 63.8 in. for women in the general population. We can conclude that supermodels are taller than typical women.

One-sample t-test / Upper-tailed test:	
99% confidence interval on the mean:	
(69.1110 , +Inf)	
Difference	6.0250
t (Observed value)	23.8237
t (Critical value)	2.8234
DF	9
p-value (one-tailed)	< 0.0001
alpha	0.01

25. Open the **QUAKE** data file on your data disk. The sample data meet the loose requirement of having a normal distribution. H_0: $\mu = 1.00$. H_1: $\mu > 1.00$. Select the **XLSTAT** add-in. Select **Parametric tests**. Select **One-sample t-test and z-test**. Enter the range of the MAG data including the label, **A1:A51**. Select **One sample, Sheet, Column labels**, and **Student's t-test**. Click the **Options** tab. Alternative hypothesis: **Mean 1 > Theoretical mean**. Theoretical mean: **1.00**. Significance level (%): **5**. Click **OK**. The output is displayed below. Test statistic [t (Observed value)] = 2.2177. t (Critical value) = 1.6766. P-value (one-tailed) = 0.0156.

Reject H_0. There is sufficient evidence to support the claim that the population of earthquakes has a mean magnitude greater than 1.00.

One-sample t-test / Upper-tailed test:	
95% confidence interval on the mean:	
(1.0449 , +Inf)	
Difference	0.1842
t (Observed value)	2.2177
t (Critical value)	1.6766
DF	49
p-value (one-tailed)	0.0156
alpha	0.05

27. Open the **FRESH15** data file on your data disk. Sort the data file by SEX so that the males' data comes before the females' data. The sample data meet the loose requirement of having a normal distribution. H_0: $\mu = 83$ kg. H_1: $\mu < 83$ kg. Select the **XLSTAT** add-in. Select **Parametric tests**. Select **One-sample t-test and z-test**. Enter the range of the males' WTSEP data including the label, **B1:B33**. Select **One sample, Sheet, Column labels**, and **Student's t-test**. Click the **Options** tab. Alternative hypothesis: **Mean 1 < Theoretical mean**. Theoretical mean: **83**. Significance level (%): **1**. Click **OK**. The output is displayed below. Test statistic [t (Observed value)] = –5.5239. t (Critical value) = –2.4528. P-value (one-tailed) < 0.0001.

Reject H_0. There is sufficient evidence to support the claim that male college students have a mean weight that is less than the 83 kg mean weight of males in the general population.

One-sample t-test / Lower-tailed test:	
99% confidence interval on the mean:	
(-Inf , 77.2840)	
Difference	-10.2813
t (Observed value)	-5.5239
t (Critical value)	-2.4528
DF	31
p-value (one-tailed)	< 0.0001
alpha	0.01

Section 8-4, Beyond the Basics

29. H_0: $\mu = 24$. H_1: $\mu < 24$. Test statistic: $z = \dfrac{19.6 - 24}{3.8/\sqrt{40}} = -7.32$. $t = -7.323$. Using Excel's NORM.S.INV function, the critical value =**NORM.S.INV(0.05)** = -1.645. Using Excel's NORM.S.DIST function, the P-value =**NORM.S.DIST(-7.323,TRUE)** = 1.21244E-13. Reject H_0. There is sufficient evidence to support the claim that Chips Ahoy reduced-fat cookies have a mean number of chocolate chips that is less than 24 (but this does not provide conclusive evidence of reduced fat).

31. H_0: $\mu = 33$ years. H_1: $\mu \neq 33$ years. Test statistic: $z = \dfrac{35.9 - 33}{11.1/\sqrt{82}} = 2.37$. Using Excel's NORM.S.INV function, the critical value =NORM.S.INV(0.005) = ±2.5758. Using Excel's NORM.S.DIST function, the P-value = **2*NORM.S.DIST-2.37, TRUE)** = 0.0178. Fail to reject H_0. There is not sufficient evidence to warrant rejection of the claim that the mean age of actresses when they win Oscars is 33 years.

33. $A = \dfrac{1.645(8 \cdot 149 + 3)}{8 \cdot 149 - 1} = 1.6505247$. The approximation yields a critical value of

$t = \sqrt{149\left(e^{1.6505247^2/149} - 1\right)} = 1.655$.

35. a. The power of 0.4274 shows that there is a 42.74% chance of supporting the claim that $\mu < 1$ W/kg when the true mean is actually 0.80 W/kg. This value of power is not very high, and it shows that the hypothesis test is not very effective in recognizing that the mean is less than 1.00 W/kg when the actual mean is 0.80 W/kg.

    b. $\beta = 0.5726$. The probability of a type II error is 0.5726. That is, there is a 0.5726 probability of making the mistake of not supporting the claim that $\mu < 1$ W/kg when in reality the population mean is 0.80 W/kg.

Section 8-5, Basic Skills and Concepts

1. a. The mean waiting time remains the same.

b. The variation among waiting times is lowered.

c. Because customers all have waiting times that are roughly the same, they experience less stress and are generally more satisfied. Customer satisfaction is improved.

d. The single line is better because it results in lower variation among waiting times, so a hypothesis test of a claim of a lower standard deviation is a good way to verify that the variation is lower with a single waiting line.

3. Use a 90% confidence interval. The conclusion based on the 90% confidence interval will be the same as the conclusion from a hypothesis test using the P-value method or the critical value method.

5. H_0: $\sigma = 0.15$ oz. H_1: $\sigma < 0.15$ oz. Test statistic: $\chi^2 = \dfrac{(36-1)0.11^2}{0.15^2} = 18.822$. Using Excel's CHISQ.INV function, the critical value **=CHISQ.INV(0.05,35)** = 22.4650. Using Excel's CHISQ.DIST function, the P-value **=CHISQ.DIST(18.822,35,TRUE)** = 0.0116. Reject H_0. There is sufficient evidence to support the claim that the population of volumes has a standard deviation less than 0.15 oz.

7. H_0: $\sigma = 0.0230$ g. H_1: $\sigma < 0.0230$ g. Test statistic: $\chi^2 = \dfrac{(37-1)0.01648^2}{0.0230^2} = 18.483$. Using Excel's CHISQ.INV function, the critical value **=CHISQ.INV(0.05,36)** = 23.2686. Using Excel's CHISQ.DIST function, the P-value **=CHISQ.DIST(18.483,36,TRUE)** = 0.0069. Reject H_0. There is sufficient evidence to support the claim that the population of weights has a standard deviation less than the specification of 0.0230 g.

9. The data appear to be from a normally distributed population. H_0: $\sigma = 10$ bpm. H_1: $\sigma \neq 10$ bpm. Test statistic: $\chi^2 = \dfrac{(40-1)10.3^2}{10^2} = 41.375$. Using Excel's functions, the critical values **=CHISQ.INV(0.025,39)** = 23.6543 and **=CHISQ.INV.RT(0.025,39)** = 58.1201. Using Excel's CHISQ.DIST.RT function, the P-value **=2*CHISQ.DIST.RT(41.375,39)** = 0.7347. Fail to reject H_0. There is not sufficient evidence to warrant rejection of the claim that pulse rates of men have a standard deviation equal to 10 beats per minute.

11. H_0: $\sigma = 3.2$ mg. H_1: $\sigma \neq 3.2$ mg. Test statistic: $\chi^2 = \dfrac{(25-1)3.7^2}{3.2^2} = 32.086$. Using Excel's functions, the critical values **=CHISQ.INV(0.025,24)** = 12.4012 and **=CHISQ.INV.RT(0.025,24)** = 39.3641. Using Excel's CHISQ.DIST.RT function, the P-value **=2*CHISQ.DIST.RT(32.086,24)** = 0.2498. Fail to reject H_0. There is not sufficient evidence to support the claim that filtered 100-mm cigarettes have tar amounts with a standard deviation different from 3.2 mg. There is not enough evidence to conclude that filters have an effect.

13. The data appear to be from a normally distributed population. H_0: $\sigma = 22.5$ years. H_1: $\sigma < 22.5$ years. Test statistic: $\chi^2 = \dfrac{(15-1)7.67^2}{22.5^2} = 1.627$. Using Excel's CHISQ.INV function, the critical value **=CHISQ.INV(0.01,14)** = 4.6604. Using Excel's CHISQ.DIST function, the P-value **=CHISQ.DIST(1.627,14,TRUE)** = 2.30545E05. Reject H_0. There is sufficient evidence to support the claim that the standard deviation of ages of all race car drivers is less than 22.5 years.

15. The data appear to be from a normally distributed population. H_0: $\sigma = 32.2$ ft. H_1: $\sigma > 32.2$ ft. Test statistic: $\chi^2 = \dfrac{(12-1)52.4^2}{32.2^2} = 29.176$. Using Excel's CHISQ.INV.RT function, the critical value **=CHISQ.INV.RT(0.05,12)** = 19.6751. Using Excel's CHISQ.DIST.RT function, the P-value **=CHISQ.DIST.RT(29.176,11)** = 0.0021. Reject H_0. There is sufficient evidence to support the claim that the new production method has errors with a standard deviation greater than 32.2 ft. The variation appears to be greater than in the past, so the new method appears to be worse, because there will be more altimeters that have larger errors. The company should take immediate action to reduce the variation.

17. The data appear to be from a normally distributed population. H_0: $\sigma = 0.15$ oz. H_1: $\sigma < 0.15$ oz. Test statistic:

$$\chi^2 = \frac{(36-1)0.0809^2}{0.15^2} = 10.173 \text{ . Using Excel's CHISQ.INV function, the critical value } = \textbf{CHISQ.INV(0.05,35)}$$

= 22.4650. Using Excel's CHISQ.DIST function, the P-value =**CHISQ.DIST(10.173,35,TRUE)** = 1.30103E-05. Reject H_0. There is sufficient evidence to support the claim that the population of volumes has a standard deviation less than 0.15 oz.

Section 8-5, Beyond the Basics

19. Critical $\chi^2 = \dfrac{1}{2}\left(-1.645 + \sqrt{2 \cdot 35 - 1}\right)^2 = 22.189$, which is reasonably close to the value of 22.4650 obtained from Excel.

Chapter Quick Quiz

1. H_0: $\mu = 0$ sec. H_1: $\mu \neq 0$ sec.

2. a. Two-tailed   b. Student t

3. a. Fail to reject H_0.

   b. There is not sufficient evidence to warrant rejection of the claim that the sample is from a population with a mean equal to 0 sec.

4. There is a loose requirement of a normally distributed population in the sense that the test works reasonably well if the departure from normality is not too extreme.

5. a. H_0: $p = 0.5$. H_1: $p > 0.5$ b. $z = \dfrac{0.64 - 0.5}{\sqrt{\frac{(0.5)(0.5)}{511}}} = 6.33$

   c. P-value = 0.0000000001263996. There is sufficient evidence to support the claim that the majority of adults are in favor of the death penalty for a person convicted of murder.

6. Using Excel, the P-value =**2*NORM.S.DIST(-2,TRUE)** = 0.0455.

7. The only true statement is the one given in part (a).

8. No. All critical values of x2 are greater than zero.

9. True

10. False

Review Exercises

1. a. False   b. True

   c. False   d. False

   e. False

2. H_0: $p = \frac{2}{3}$. H_1: $p \neq \frac{2}{3}$. Test statistic: $z = \dfrac{\frac{657}{1010} - \frac{2}{3}}{\sqrt{\frac{\left(\frac{2}{3}\right)\left(\frac{1}{3}\right)}{1010}}} = -1.09$. Using Excel's NORM.S.INV function, the critical

   values are =**NORM.S.INV(0.005)** = -2.576 and =**NORM.S.INV(0.995)** = 2.576. Using Excel's NORM.S.DIST function, the P-value =**2*NORM.S.DIST(-1.09, TRUE)** = 0.2757. Fail to reject H_0. There is not sufficient evidence to warrant rejection of the claim that 2/3 of adults are satisfied with the amount of leisure time that they have.

3. H_0: $p = 0.75$. H_1: $p > 0.75$. Test statistic: $z = \dfrac{\frac{678}{737} - 0.75}{\sqrt{\frac{(0.75)(0.25)}{737}}} = 10.65$ or $z = 10.66$ (if using $\hat{p} = 0.92$). Using

Excel's NORM.S.INV function, the critical value is =**NORM.S.INV(0.99)** = 2.3263. Using Excel's
NORM.S.DIST function, the P-value =**NORM.S.DIST(-10.65, TRUE)** = 8.71825E-27. Reject H_0. There is
sufficient evidence to support the claim that more than 75% of us do not open unfamiliar e-mail and instant-
message links. Given that the results are based on a voluntary response sample, the results are not necessarily
valid.

4. H_0: $\mu = 3369$ g. H_1: $\mu < 3369$ g. Test statistic: $t = \dfrac{3245 - 567}{446/\sqrt{81}} = -19.962$. Using Excel's NORM.S.INV

function, the critical value is =**NORM.S.INV(0.01)** = -2.3263. Using Excel's NORM.S.DIST function, the P-
value =**NORM.S.DIST(-19.962, TRUE)** = 5.8949E-89. Reject H_0. There is sufficient evidence to support the
claim that the mean birth weight of Chinese babies is less than the mean birth weight of 3369 g for Caucasian
babies.

5. H_0: $\sigma = 567$ g. H_1: $\sigma \neq 567$ g. Test statistic: $\chi^2 = \dfrac{(81-1)\,466^2}{567^2} = 54.038$. Using Excel's functions, the critical

values =**CHISQ.INV(0.005,80)** = 51.1719 and =**CHISQ.INV.RT(0.005,80)** = 116.3211. Using Excel's
CHISQ.DIST function, the P-value =**2*CHISQ.DIST(54.038,80)** = 0.0229. Fail to reject H_0. There is not
sufficient evidence to warrant rejection of the claim that the standard deviation of birth weights of Chinese
babies is equal to 567 g.

6. H_0: $\mu = 1.5$ ug/m^3. H_1: $\mu > 1.5$ mg/m^3. Enter the label **ug/m^3** in cell A1 of an Excel worksheet followed by
the data given in Exercise 6. Select the **XLSTAT** add-in. Select **Parametric tests**. Select **One-sample t-test
and z-test**. Enter the data range including the label, **A1:A7**. Select **One sample**, **Sheet**, **Column labels**, and
Student's t-test. Click the **Options** tab. Alternative hypothesis: **Mean 1 > Theoretical mean**. Theoretical
mean: **1.5**. Significance level (%): **5**. Click **OK**. The output is displayed below. Test statistic [t (Observed
value)] = 0.0491. t (Critical value) = 3.3650. P-value (one-tailed) = 0.4814.

Fail to reject H_0. There is not sufficient evidence to support the claim that the sample is from a population with
a mean greater than the EPA standard of 1.5 mg/m^3.

One-sample t-test / Upper-tailed test:		
99% confidence interval on the mean:		
(-1.0913 , +Inf)		
Difference	0.0383	
t (Observed value)	0.0491	
t (Critical value)	3.3650	
DF	5	
p-value (one-tailed)	0.4814	
alpha	0.01	

To construct a normal quantile plot, select the **XLSTAT** add-in. Select **Describing data**. Select **Normality
tests**. Enter the data range, **A1:A7**. Select **Sheet** and **Sample labels**. Set the Significance level (%) equal to **5**.
Click the **Charts** tab. Select **Normal Q-Q plots**. Click **OK**. Because the sample value of 5.40 mg/m^3 appears to
be an outlier and because a normal quantile plot suggests that the sample data are not from a normally
distributed population, the requirements of the hypothesis test are not satisfied, and the results of the hypothesis
test are therefore questionable.

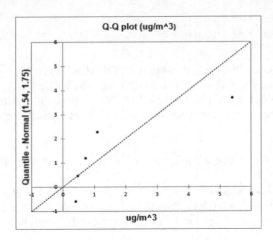

7. H_0: $\mu = 25$. H_1: $\mu \neq 25$. Test statistic: $t = \dfrac{24.2 - 25}{14.1/\sqrt{100}} = -0.567$. Using Excel's T.INV.2T function, the

critical value =**T.INV.2T(0.01,99)** $= \pm 2.6264$. Using Excel's T.DIST.2T function, the P-value
=**T.DIST.2T(0.567,99)** $= 0.5720$. Fail to reject H_0. There is not sufficient evidence to warrant rejection of the
claim that the sample is selected from a population with a mean equal to 25.

8.   a.   A type I error is the mistake of rejecting a null hypothesis when it is actually true. A type II error is the
          mistake of failing to reject a null hypothesis when in reality it is false.

      b.   Type I error: Reject the null hypothesis that the mean of the population is equal to 25 when in reality, the
          mean is actually equal to 25. Type II error: Fail to reject the null hypothesis that the population mean is
          equal to 25 when in reality, the mean is actually different from 25.

9.   The χ^2 test has a reasonably strict requirement that the sample data must be randomly selected from a
      population with a normal distribution, but the numbers are selected in such a way that they are all equally likely,
      so the population has a uniform distribution instead of the required normal distribution. Because the
      requirements are not all satisfied, the χ^2 2 test should not be used.

10. The sample data meet the loose requirement of having a normal distribution. H_0: $\mu = 1000$ HIC.

      H_1: $\mu < 1000$ HIC. Enter the label **HIC** in cell A1 of an Excel worksheet followed by the data given in Exercise
      10. Select the **XLSTAT** add-in. Select **Parametric tests**. Select **One-sample *t*-test and *z*-test**. Enter the data
      range including the label, **A1:A6**. Select **One sample, Sheet, Column labels**, and **Student's *t*-test**. Click the
      Options tab. Alternative hypothesis: **Mean 1 < Theoretical mean**. Theoretical mean: **1000**. Significance level
      (%): **1**. Click **OK**. The output is displayed below. Test statistic [t (Observed value)] $= -10.1769$. t (Critical
      value) $= -3.7470$. P-value (one-tailed) $= 0.0003$. Reject H_0. There is sufficient evidence to support the claim
      that the population mean is less than 1000 HIC. The results suggest that the population mean is less than 1000
      HIC, so they appear to satisfy the specified requirement.

One-sample t-test / Lower-tailed test:	
99% confidence interval on the mean:	
(-Inf , 781.2648)	
Difference	-346.2000
t (Observed value)	-10.1769
t (Critical value)	-3.7470
DF	4
p-value (one-tailed)	0.0003
alpha	0.01

Cumulative Review Exercises

1. Enter the data in an Excel worksheet. Click on **DATA**. Select **Data Analysis** and select **Descriptive Statistics**. Enter the data range and select **Summary statistics**. Select Confidence Level for Mean: **95%**. Click **OK**.

   a. The mean is 53.3 words.   b. The median is 52 words.

   c. The standard deviation is 15.7 words.   d. The variance is 245.1 words squared.

   e. The range is 45 words.

Number of Words	
Mean	53.3
Standard Error	4.95098
Median	52
Mode	#N/A
Standard Deviation	15.6564
Sample Variance	245.122
Kurtosis	-1.1323
Skewness	0.35449
Range	45
Minimum	34
Maximum	79
Sum	533
Count	10
Confidence Level(95.0%)	11.1999

2. a. Ratio   b.   Discrete

   c. The sample is a simple random sample if it was selected in such a way that all possible samples of the same size have the same chance of being selected.

3. Use the confidence level value from the Excel output for Exercise 1 to construct the 95% confidence interval.

   53.3-11.1999 words $< \mu <$ 53.3+11.1999 words

   42.1 words $< \mu <$ 64.5 words

4. H_0: $\mu = 48.0$ words. H_1: $\mu > 48.0$ words. Test statistic: $t = \dfrac{53.3 - 48.0}{15.7/\sqrt{10}} = 1.070$. Using Excel's T.INV

   function, the critical value **=T.INV0.95,9)** = 1.8331. Using Excel's T.DIST function, the *P*-value **=1-T.DIST(1.070, 9)** = 0.1562. Fail to reject H_0. There is not sufficient evidence to support the claim that the mean number of words on a page is greater than 48.0. There is not enough evidence to support the claim that there are more than 70,000 words in the dictionary.

5. a. $z = \dfrac{38.8 - 36.0}{1.4} = 2$; Using Excel's NORM.S.DIST function, $P(z > 2)$ **=1 – NORM.S.DIST(2,TRUE)** = 0.02275. $P(z > 2) = 2.28\%$.

   b. Use Excel's NORM.S.INV function to find the z-score that corresponds to the 98th percentile, **=NORM.S.INV(0.98)** = 2.0537. 98th percentile: $x = \mu + z \cdot \sigma = 36.0 + 2.0537 \cdot 1.4 = 38.9$ in.

   c. $z = \dfrac{37.0 - 36.0}{1.4/\sqrt{4}} = 1.43$; Using Excel's NORM.S.DIST function,

   $P(z < 1.43)$ **=NORM.S.DIST(1.43,TRUE)** = 0.9236. $P(z < 1.43) = 92.36\%$.

6. a. $(0.125)^3 = 0.00195$. It is unlikely because the probability of the event occurring is so small.

    b. $(0.097)(0.125) = 0.0121$

    c. $1 - (0.875)^5 = 0.487$

7. No. The distribution is very skewed. A normal distribution would be approximately bell-shaped, but the displayed distribution is very far from being bell-shaped.

8. Because the vertical scale starts at 7000 and not at 0, the difference between the number of males and the number of females is exaggerated, so the graph is deceptive by creating the wrong impression that there are many more male graduates than female graduates.

9. a. $0.372(1003) = 373$

    b. $\hat{p} \pm z_{\alpha/2} \sqrt{\dfrac{\hat{p}\hat{q}}{n}} = 0.372 \pm 1.96 \sqrt{\dfrac{(0.372)(0.628)}{1003}}$

    $0.342 < p < 0.402$ or $34.2\% < p < 40.2\%$.

    c. $z = \dfrac{\hat{p} - p}{\sqrt{\dfrac{pq}{n}}} = \dfrac{0.372 - 0.5}{\sqrt{\dfrac{(0.5)(0.5)}{1003}}} = -8.1085$.

    Using Excel's NORM.S.DIST function, the P-value =**NORM.S.DIST(-8.1085,TRUE)** = 2.5624E-16. With test statistic $z = -8.1085$ and with a P-value close to 0, there is sufficient evidence to support the claim that less than 50% of adults answer "yes."

    d.   The required sample size depends on the confidence level and the sample proportion, not the population size.

10. H_0: $p = 0.5$. H_1: $p < 0.5$. Test statistic: $z = \dfrac{0.372 - 0.5}{\sqrt{\dfrac{(0.5)(0.5)}{1003}}} = -8.1085$. Using Excel's NORM.S.INV function,

    the critical value =NORM.S.INV(0.01) = –2.3263. Using Excel's NORM.S.DIST function, the P-value =**NORM.S.DIST(-8.1085,TRUE)** = 2.5624E-16. Reject H_0. There is sufficient evidence to support the claim that fewer than 50% of Americans say that they have a gun in their home.

Chapter 9

Inferences from Two Samples

Section 9-2, Basic Skills and Concepts

1. The samples are simple random samples that are independent. For each of the two groups, the number of successes is at least 5 and the number of failures is at least 5. (Depending on what we call a success, the four numbers are 33, 115, 201,229 and 200,745 and all of those numbers are at least 5.) The requirements are satisfied.

3. a. H_0: $p_1 = p_2$. H_1: $p_1 < p_2$.

   b. If the P-value is less than 0.001 we should reject the null hypothesis and conclude that there is sufficient evidence to support the claim that the rate of polio is less for children given the Salk vaccine than it is for children given a placebo.

5. Test statistic: $z = -12.39$ (rounded). The P-value is < 0.0001. There is sufficient evidence to warrant rejection of the claim that the vaccine has no effect.

For Exercises 7 – 17, assume that the data fit the requirements for the statistical methods for two proportions unless otherwise indicated.

7. a. H_0: $p_1 = p_2$. H_1: $p_1 > p_2$. Test statistic: $z = 6.44$. Using Excel's NORM.S.INV function, the critical value **=NORM.S.INV(0.99)** = 2.3263. Using Excel's NORM.S.DIST function, the P-value **=1-NORM.S.DIST(6.44,TRUE)** = 5.97368E-11. Reject H_0. There is sufficient evidence to support the claim that the proportion of people over 55 who dream in black and white is greater than the proportion for those under 25.

   b. 98% CI: $0.117 < p_1 - p_2 < 0.240$. Because the confidence interval limits do not include 0, it appears that the two proportions are not equal. Because the confidence interval limits include only positive values, it appears that the proportion of people over 55 who dream in black and white is greater than the proportion for those under 25.

   c. The results suggest that the proportion of people over 55 who dream in black and white is greater than the proportion for those under 25, but the results cannot be used to verify the cause of that difference.

9. a. H_0: $p_1 = p_2$. H_1: $p_1 > p_2$. Test statistic: $z = 6.11$. Using Excel's NORM.S.INV function, the critical value **=NORM.S.INV(0.95)** = 1.645. Using Excel's NORM.S.DIST function, the P-value **=1-NORM.S.DIST(6.11,TRUE)** = 4.98156E-10. Reject H_0. There is sufficient evidence to support the claim that the fatality rate is higher for those not wearing seat belts.

   b. 90% CI: $0.00556 < p_1 - p_2 < 0.0122$. Because the confidence interval limits do not include 0, it appears that the two fatality rates are not equal. Because the confidence interval limits include only positive values, it appears that the fatality rate is higher for those not wearing seat belts.

   c. The results suggest that the use of seat belts is associated with lower fatality rates than not using seat belts.

11. a. H_0: $p_1 = p_2$. H_1: $p_1 \neq p_2$. Test statistic: $z = 0.57$. Using Excel's NORM.S.INV function, the critical values **=NORM.S.INV(0.975)** = ± 1.9600. Using Excel's NORM.S.DIST function, the P-value **=2*(1-NORM.S.DIST(0.57,TRUE))** = 0.5687. Fail to reject H_0. There is not sufficient evidence to support the claim that echinacea treatment has an effect.

    b. 95% CI: $-0.0798 < p_1 - p_2 < 0.149$. Because the confidence interval limits do contain 0, there is not a significant difference between the two proportions. There is not sufficient evidence to support the claim that echinacea treatment has an effect.

    c. Echinacea does not appear to have a significant effect on the infection rate. Because it does not appear to have an effect, it should not be recommended.

13. a. H_0: $p_1 = p_2$. H_1: $p_1 \neq p_2$. Test statistic: $z = 0.40$. Using Excel's NORM.S.INV function, the critical values =**NORM.S.INV(0.975)** = ±1.9600. Using Excel's NORM.S.DIST function, the P-value =**2*(1-NORM.S.DIST(0.40,TRUE))** = 0.6892. Fail to reject H_0. There is not sufficient evidence to warrant rejection of the claim that men and women have equal success in challenging calls.

   b.   95% CI: $-0.0318 < p_1 - p_2 < 0.0484$. Because the confidence interval limits contain 0, there is not a significant difference between the two proportions. There is not sufficient evidence to warrant rejection of the claim that men and women have equal success in challenging calls.

   c.   It appears that men and women have equal success in challenging calls.

15. a. H_0: $p_1 = p_2$. H_1: $p_1 > p_2$. Test statistic: $z = 9.97$. Using Excel's NORM.S.INV function, the critical value =**NORM.S.INV(0.99)** = 2.3264. Using Excel's NORM.S.DIST function, the P-value =**1-NORM.S.DIST(9.97,TRUE)** = 0.0000. Reject H_0. There is sufficient evidence to support the claim that the cure rate with oxygen treatment is higher than the cure rate for those given a placebo. It appears that the oxygen treatment is effective.

   b.   98% CI: $0.467 < p_1 - p_2 < 0.687$. Because the confidence interval limits do not include 0, it appears that the two cure rates are not equal. Because the confidence interval limits include only positive values, it appears that the cure rate with oxygen treatment is higher than the cure rate for those given a placebo. It appears that the oxygen treatment is effective.

   c.   The results suggest that the oxygen treatment is effective in curing cluster headaches.

17. a. H_0: $p_1 = p_2$. H_1: $p_1 < p_2$. Test statistic: $z = -1.17$. Using Excel's NORM.S.INV function, the critical value =**NORM.S.INV(0.01)** = -2.3264. Using Excel's NORM.S.DIST function, the P-value =**NORM.S.DIST(-1.17,TRUE)** = 0.1210. Fail to reject H_0. There is not sufficient evidence to support the claim that the rate of left-handedness among males is less than that among females.

   b.   98% CI: $-0.0849 < p_1 - p_2 < 0.0265$. Because the confidence interval limits include 0, there does not appear to be a significant difference between the rate of left-handedness among males and the rate among females. There is not sufficient evidence to support the claim that the rate of left-handedness among males is less than that among females.

   c.   The rate of left-handedness among males does not appear to be less than the rate of left-handedness among females.

Section 9-2, Beyond the Basics

19. a. $0.0227 < p_1 - p_2 < 0.217$; because the confidence interval limits do not contain 0, it appears that H_0: $p_1 = p_2$ can be rejected.

   b. $0.491 < p_1 < 0.629$; $0.371 < p_2 < 0.509$; because the confidence intervals do overlap, it appears that H_0: $p_1 = p_2$ cannot be rejected.

   c. H_0: $p_1 = p_2$. H_1: $p_1 \neq p_2$. Test statistic: $z = 2.40$. Using Excel's NORM.S.INV function, the critical values =**NORM.S.INV(0.975)** = ±1.9600. Using Excel's NORM.S.DIST function, the P-value =**2*(1-NORM.S.DIST(2.40,TRUE))** = 0.0164. Reject H_0. There is sufficient evidence to reject H_0: $p_1 = p_2$.

   d.   Reject H_0: $p_1 = p_2$. Least effective: Using the overlap between the individual confidence intervals.

21. $n = \dfrac{z_{\alpha/2}^2}{2E^2} = \dfrac{1.645^2}{2 \cdot 0.02^2} = 3383$

Section 9-3, Basic Skills and Concepts

1. Independent: b, d, e

3. Because the confidence interval does not contain 0, it appears that there is a significant difference between the mean height of women and the mean height of men. Based on the confidence interval, it appears that the mean height of men is greater than the mean height of women.

5. a. $H_0: \mu_1 = \mu_2$. $H_1: \mu_1 \neq \mu_2$. Test statistic: $t = -2.979$. $df = 34$. Using Excel's T.INV.2T function, the critical values =**T.INV.2T(0.05,34)** = ± 2.0322. Using Excel's T.DIST.2T function, the P-value =**T.DIST.2T(2.979,34)** = 0.0053. Reject H_0. There is sufficient evidence to warrant rejection of the claim that the samples are from populations with the same mean. Color does appear to have an effect on creativity scores. Blue appears to be associated with higher creativity scores.

   b. $E = 0.3957$; 95% CI: $-0.98 < \mu_1 - \mu_2 < -0.18$

7. a. $H_0: \mu_1 = \mu_2$. $H_1: \mu_1 > \mu_2$. Test statistic: $t = 0.132$. $df = 19$. Using Excel's T.INV function, the critical value =**T.INV(0.95,19)** = 1.7291. Using Excel's T.DIST.RT function, the P-value =**T.DIST.RT(0.132,19)** = 0.4482. Fail to reject H_0. There is not sufficient evidence to support the claim that the magnets are effective in reducing pain. It is valid to argue that the magnets might appear to be effective if the sample sizes are larger.

   b. $E = 0.6563$; 90% CI: $-0.61 < \mu_1 - \mu_2 < 0.71$

9. a. **Note: At the time this solutions manual was being prepared, the headings were reversed in the Table for Exercise 9 displayed in the textbook.** The correct table values are as follows. Women: $n_1 = 11$, $\overline{x}_1 = 97.69°$ F, $s_1 = 0.89°$F. Men: $n_2 = 59$, $\overline{x}_1 = 97.45°$ F, $s_2 = 0.66°$F. In the solution given here, females comprise group 1 and males comprise group 2. $H_0: \mu_1 = \mu_2$. $H_1: \mu_1 > \mu_2$. Test statistic: $t = 0.852$. $df = 10$. Using Excel's T.INV function, the critical value =**T.INV(0.99,10)** = 2.7638. Using Excel's T.DIST.RT function, the P-value =**T.DIST.RT(0.0.852,10)** = 0.2071. Fail to reject H_0. There is not sufficient evidence to support the claim that women have a higher mean body temperature than men.

   b. $E = 0.7787$; 98% CI: $-0.54 °F < \mu_1 - \mu_2 < 1.02 °F$

11. a. $H_0: \mu_1 = \mu_2$. $H_1: \mu_1 < \mu_2$. Test statistic: $t = -3.547$. $df = 29$. Using Excel's T.INV function, the critical value =**T.INV(0.01,29)** = -2.4620. Using Excel's T.DIST function, the P-value =**T.DIST(-3.547,29,TRUE)** = 0.0007. Reject H_0. There is sufficient evidence to support the claim that the mean maximal skull breadth in 4000 b.c. is less than the mean in a.d. 150.

    b. $E = 3.33118$; 98% CI: -8.13 mm $< \mu_1 - \mu_2 < -1.47$ mm

13. a. $H_0: \mu_1 = \mu_2$. $H_1: \mu_1 < \mu_2$. Test statistic: $t = -3.142$. $df = 29$. Using Excel's T.INV function, the critical value =**T.INV(0.01,29)** = -2.4620. Using Excel's T.DIST function, the P-value =**T.DIST(-3.142,29,TRUE)** = 0.0019. Reject H_0. There is sufficient evidence to support the claim that students taking the nonproctored test get a higher mean than those taking the proctored test.

    b. $E = 11.2205$; 98% CI: $-25.54 < \mu_1 - \mu_2 < -3.10$

15. a. $H_0: \mu_1 = \mu_2$. $H_1: \mu_1 \neq \mu_2$. Test statistic: $t = 1.274$. $df = 39$. Using Excel's T.INV.2T function, the critical value =**T.INV.2T(0.05,39)** = ± 2.0227. Using Excel's T.DIST.2T function, the P-value =**T.DIST.2T(1.274,39)** = 0.2102. Fail to reject H_0. There is not sufficient evidence to warrant rejection of the claim that males and females have the same mean BMI.

    b. $E = 2.9206$; 95% CI: $-1.08 < \mu_1 - \mu_2 < 4.76$

17. a. $H_0: \mu_1 = \mu_2$. $H_1: \mu_1 > \mu_2$. Open the **IQLEAD** data file on your data disk. Click **DATA** at the top of the screen. Select **Data Analysis**. Select **t-Test: Two-Sample Assuming Unequal Variances**. Click **OK**. Enter the IQF data range for LEAD group 2 (**H80:H101**) in the variable 1 range window. Enter the IQF data range for LEAD group 3 (**H102:H122**) in the variable 2 range window. The hypothesized mean

difference = **0**. Alpha = **0.05**. Select **New Worksheet Ply**. Click **OK**. The output is shown below. Test statistic (t Stat): $t = 0.0890$. Critical value (*t* Critical one-tail): $t = 1.6883$. *P*-value [$P(T<=t)$ one-tail] = 0.4648. Fail to reject H_0. There is not sufficient evidence to support the claim that the mean IQ score of people with medium lead levels is higher than the mean IQ score of people with high lead levels.

t-Test: Two-Sample Assuming Unequal Variances

	Variable 1	Variable 2
Mean	87.22727273	86.9047619
Variance	204.2792208	80.79047619
Observations	22	21
Hypothesized Mean Difference	0	
df	36	
t Stat	0.088995734	
P(T<=t) one-tail	0.464789284	
t Critical one-tail	1.688297714	
P(T<=t) two-tail	0.929578568	
t Critical two-tail	2.028094001	

   b. $E = 6.1182$; 90% CI: $-5.8 < \mu_1 - \mu_2 < 6.4$

19. a. $H_0: \mu_1 = \mu_2$. $H_1: \mu_1 < \mu_2$. Enter the label **Popes** in cell A1 of an Excel worksheet followed by the popes' longevity data given in Exercise 19. Enter the label **Kings & Queens** in cell B1 followed by the monarchs' longevity data given in Exercise 19. Click **DATA** at the top of the screen. Select **Data Analysis**. **Select t-Test: Two-Sample Assuming Unequal Variances**. Click **OK**. Enter the popes' longevity data range (**A1:A25**) in the variable 1 range window. Enter the kings' and queens' longevity data range (**B1:B15**) in the variable 2 range window. The hypothesized mean difference = **0**. Select **Labels**. Alpha = **0.01**. Click **OK**. The output is displayed below. Test statistic (*t* Stat): $t = -1.8101$. Critical value (*t* Critical one-tail): $t = -2.5669$. *P*-value [$P(T<=t)$ one-tail] = 0.0440. Fail to reject H_0. There is not sufficient evidence to support the claim that the mean longevity for popes is less than the mean for British monarchs after coronation.

t-Test: Two-Sample Assuming Unequal Variances

	Popes	Kings & Queens
Mean	13.125	22.71428571
Variance	80.28804348	346.0659341
Observations	24	14
Hypothesized Mean Difference	0	
df	17	
t Stat	-1.810126457	
P(T<=t) one-tail	0.043994571	
t Critical one-tail	2.566933984	
P(T<=t) two-tail	0.087989141	
t Critical two-tail	2.89823052	

   b. $E = 13.5985$; 98% CI: -23.2 years $< \mu_1 - \mu_2 < 4.0$ years

21. $H_0: \mu_1 = \mu_2$. $H_1: \mu_1 \neq \mu_2$. Open the **COINS** data file on the data disk. Select the **XLSTAT** add-in. Select **Parametric tests**. Select **Two-sample *t*-test and *z*-test**. Enter the range of the Pre-1964 data (**F1:F41**) in the sample 1 window. Enter the range of the Post-1964 data (**G1:G41**) in the sample 2 window. Select **One column per sample**. Select **Column labels**. Select **Student's *t*-test**. Click the **Options** tab. Alternative hypothesis: **Mean 1 – Mean 2 ≠ D**. Hypothesized difference (D): **0**. Significance level (%): **5**. Click **OK**. The output is displayed at the top of the next page. Test statistic [*t* (Observed value)]: $t = 32.7729$. *t* Critical values: $t = \pm1.9942$. *P*-value (Two-tailed) < 0.0001. Reject H_0. There is sufficient evidence to warrant rejection of the claim that the two populations have equal means. The difference is highly significant, even though the samples are relatively small.

Summary statistics:

Variable	Observations	Obs. with missing data	Obs. without missing data	Minimum	Maximum	Mean	Std. deviation
Pre-1964 Quarters	40	0	40	6.0002	6.3669	6.1927	0.0870
Post-1964 Quarters	40	0	40	5.5361	5.7790	5.6393	0.0619

t-test for two independent samples / Two-tailed test:

95% confidence interval on the difference between the means:
(0.5197, 0.5870)

Difference	0.5534
t (Observed value)	32.7729
|t| (Critical value)	1.9942
DF	70
p-value (Two-tailed)	< 0.0001
alpha	0.05

The number of degrees of freedom is approximated by the Welch-Satterthwaite formula

23. Open the **COLA** data file on your data disk. Select the **XLSTAT** add-in. Select **Parametric tests**. Select **Two-sample *t*-test and *z*-test**. Enter the range of the PPREGWT data (**E1:E37**) in the sample 1 window. Enter the range of the PPDIETWT data (**G1:G37**) in the sample 2 window. Select **One column per sample**. Select **Column labels**. Select **Student's *t*-test**. Click the **Options** tab. Alternative hypothesis: **Mean 1 – Mean 2 ≠ D**. Hypothesized difference (D): **0**. Significance level (%): **5**. Click **OK**. The output is displayed below. 95% CI: 0.0379 lb $< \mu_1 - \mu_2 < 0.0426$. Because the confidence interval does not include 0, there appears to be a significant difference between the two population means. It appears that the cola in cans of regular Pepsi weighs more than the cola in cans of Diet Pepsi, and that is probably due to the sugar in regular Pepsi that is not in Diet Pepsi.

Summary statistics:

Variable	Observations	Obs. with missing data	Obs. without missing data	Minimum	Maximum	Mean	Std. deviation
PPREGWT	36	0	36	0.8139	0.8401	0.8241	0.0057
PPDIETWT	36	0	36	0.7742	0.7938	0.7839	0.0044

t-test for two paired samples / Two-tailed test:

95% confidence interval on the difference between the means:
(0.0379, 0.0426)

Difference	0.0402
t (Observed value)	34.3936
|t| (Critical value)	2.0301
DF	35
p-value (Two-tailed)	< 0.0001
alpha	0.05

Section 9-3, Beyond the Basics

25. a. **Note: At the time this solutions manual was being prepared, the headings were reversed in the Table for Exercise 9 displayed in the textbook.** The correct table values are as follows. Women: $n_1 = 11$, $\overline{x}_1 = 97.69°$ F, $s_1 = 0.89°$F. Men: $n_2 = 59$, $\overline{x}_1 = 97.45°$ F, $s_2 = 0.66°$F. In the solution given here, females comprise group 1 and males comprise group 2. $H_0: \mu_1 = \mu_2$. $H_1: \mu_1 > \mu_2$. Test statistic: $t = 1.046$. Using Excel's T.INV function, the critical value = T.INV(0.99,68) = 2.3824. Using Excel's T.DIST.RT function, the *P*-value =**T.DIST.RT(1.046,68)** = 0.1496. Fail to reject H_0. There is not sufficient evidence to support the claim that women have a higher mean body temperature than men.

b. $E = 0.6713$; 98% CI: $-0.43°F < \mu_1 - \mu_2 < 0.91°F$.

27. H_0: $\mu_1 = \mu_2$. H_1: $\mu_1 \neq \mu_2$. Test statistic: $t = 15.322$. Critical values: $t = \pm 2.0180$. P-value = 8.23243E-19. Reject H_0. There is sufficient evidence to warrant rejection of the claim that the two populations have the same mean.

$$t = \frac{(0.049 - 0.000) - (\mu_1 - \mu_2)}{\sqrt{\dfrac{s_p^2}{22} + \dfrac{s_p^2}{22}}}; \quad s_p^2 = \frac{(22-1)0.015^2 + (22-1)0^2}{(22-1)+(22-1)} = 0.000125$$

29. a. H_0: $\mu_1 = \mu_2$. H_1: $\mu_1 < \mu_2$. Test statistic: $t = -3.002$. Critical t value based on 68.9927614 degrees of freedom: =**T.INV(0.01,68.9927614)** = -2.3824. P-value =**T.DIST(-3.002,68.9927614,TRUE)** = 0.0019. Reject H_0. There is sufficient evidence to support the claim that students taking the nonproctored test get a higher mean than those taking the proctored test.

b. $-25.69 < \mu_1 - \mu_2 < -2.95$

Section 9-4, Basic Skills and Concepts

1.   Parts (c) and (e) are true.

3.   The test statistic will remain the same. The confidence interval limits will be expressed in the equivalent values of km/L.

5. H_0: $\mu_d = 0$ cm. H_1: $\mu_d > 0$ cm. Open the **POTUS** data file on your data disk. Select the **XLSTAT** add-in. Select **Parametric tests**. Select **Two-sample *t*-test and *z*-test**. Enter the range of the presidents' heights (**E1:E39**) in the sample 1 window. Enter the range of the opponents' heights (**F1:F39**) in the sample 2 window. Select **Paired samples**. Select **Column labels**. Select **Student's *t* test**. Click the **Options** tab. Alternative hypothesis: **Mean 1 – Mean 2 > D**. Hypothesized difference (D): **0**. Significance level (%): **5**. Click the **Missing data** tab. Select **Remove the observations**. Click **OK**. The output is displayed below. Test statistic [t (Observed value)]: t = 0.0357. t (Critical value): t = 1.6924. P-value (one-tailed) = 0.4859. Fail to reject H_0. There is not sufficient evidence to support the claim that for the population of heights of presidents and their main opponents; the differences have a mean greater than 0 cm (with presidents tending to be taller than their opponents).

Summary statistics:

Variable	Observations	Obs. with missing data	Obs. without missing data	Minimum	Maximum	Mean	Std. deviation
Ht	34	0	34	168.0000	193.0000	180.0294	6.8953
HtOpp	34	0	34	168.0000	196.0000	179.9706	6.2011

t-test for two paired samples / Upper-tailed test:

95% confidence interval on the difference between the means:
(-2.7286 , +Inf)

Difference	0.0588
t (Observed value)	0.0357
t (Critical value)	1.6924
DF	33
p-value (one-tailed)	0.4859
alpha	0.05

7. a. $\bar{d} = -11.6$ years

b. $s_d = 17.2$ years

c.   Test statistic $t = \dfrac{\bar{d} - \mu_d}{s_d / \sqrt{n}} = \dfrac{-11.6 - 0}{17.2 / \sqrt{5}} = -1.508$

d. H_0: $\mu_d = 0$. H_1: $\mu_d \neq 0$. Using Excel's T.INV.2T function, the t critical values =**T.INV.2T**(0.05,4) = ±2.7764.

9. H_0: $\mu_d = 0$. H_1: $\mu_d \neq 0$. Test statistic: $t = \dfrac{-11.6 - 0}{17.21/\sqrt{5}} = -1.508$. Using Excel's T.INV.2T function, the t critical

values =**T.INV.2T**(0.05,4) = ±2.7764. Using Excel's T.DIST.2T function, the P-value =**T.DIST.2T(1.508,4)** = 0.2060. Fail to reject H_0. There is not sufficient evidence to support the claim that there is a difference between the ages of actresses and actors when they win Oscars.

11. Enter the label **TAXI-OUT TIME** in cell A1 of an Excel worksheet followed by the data given in the exercise. Enter the label **TAXI-IN TIME** in cell B1 followed by the data given in the exercise. Select the **XLSTAT** add-in. Select **Parametric tests**. Select **Two-sample t-test and z-test**. Enter the range of the taxi-out times, including the label (**A1:A13**), in the sample 1 window. Enter the range of the taxi-in times, including the label (**B1:B13**) in the sample 2 window. Select **Paired samples**. Select **Column labels**. Select **Student's t test**. Click the **Options** tab. Alternative hypothesis: **Mean 1 – Mean 2 ≠ D**. Hypothesized difference (D): **0**. Significance level (%): **10**. Click **OK**. The output is displayed below. The 90% CI is 1.0 min $< \mu_d <$ 12.0 min. Because the confidence interval includes only positive values and does not include 0 min, it appears that the taxi-out times are greater than the corresponding taxi-in times, so there is sufficient evidence to support the claim of the flight operations manager that for flight delays, more of the blame is attributable to taxi-out times at JFK than taxi-in times at LAX.

Summary statistics:

Variable	Observation	Obs. with missing data	Obs. without missing data	Minimum	Maximum	Mean	Std. deviation
TAXI-OUT TIME	12	0	12	12.0000	37.0000	20.4167	8.0731
TAXI-IN TIME	12	0	12	8.0000	26.0000	13.9167	5.1779

t-test for two paired samples / Two-tailed test:

90% confidence interval on the difference between the means:
(0.9890, 12.0110)

Difference	6.5000
t (Observed value)	2.1182
|t| (Critical value)	1.7959
DF	11
p-value (Two-tailed)	0.0578
alpha	0.1

13. H_0: $\mu_d = 0$. H_1: $\mu_d > 0$. Enter the label **MALE** in cell A1 of an Excel worksheet followed by the males' data given in the exercise. Enter the label **FEMALE** in cell B1 followed by the females' data given in the exercise. Select the **XLSTAT** add-in. Select **Parametric tests**. Select **Two-sample t-test and z-test**. Enter the range of the males' data, including the label (**A1:A7**), in the sample 1 window. Enter the range of the females' data, including the label (**B1:B7**) in the sample 2 window. Select **Paired samples**. Select **Column labels**. Select **Student's t test**. Click the **Options** tab. Alternative hypothesis: **Mean 1 - Mean 2 > D**. Hypothesized difference (D): **0**. Significance level (%): **5**. Click **OK**. The output is displayed at the top of the next page. Test statistic: t (Observed value) = 2.5789. t (Critical value) = 2.0158. P-value (one-tailed) = 0.0247. Reject H_0. There is sufficient evidence to support the claim that among couples, males speak more words in a day than females.

Summary statistics:

Variable	Observations	Obs. with missing data	Obs. without missing data	Minimum	Maximum	Mean	Std. deviation
MALE	6	0	6	5638.0000	46978.0000	23963.5000	13586.5111
FEMALE	6	0	6	5198.0000	31553.0000	16683.3333	8962.4343

t-test for two paired samples / Upper-tailed test:

95% confidence interval on the difference between the means:

(1589.6021, +Inf)

Difference	7280.1667
t (Observed value)	2.5789
t (Critical value)	2.0158
DF	5
p-value (one-tailed)	0.0247
alpha	0.05

15. Enter the label **6TH** in cell A1 of an Excel worksheet followed by the Friday the 6th data given in the exercise. Enter the label **13TH** in cell B1 followed by the Friday the 13th data given in the exercise. Select the **XLSTAT** add-in. Select **Parametric tests**. Select **Two-sample *t*-test and *z*-test**. Enter the range of the Friday the 6th data, including the label (**A1:A7**), in the sample 1 window. Enter the range of the Friday the 13th data, including the label (**B1:B7**) in the sample 2 window. Select **Paired samples**. Select **Column labels**. Select **Student's *t* test**. Click the **Options** tab. Alternative hypothesis: **Mean 1 - Mean 2 ≠ D**. Hypothesized difference (D): **0**. Significance level (%): **5**. Click **OK**. The 95% CI is $-6.5 < \mu_d < -0.2$. Because the confidence interval does not include 0, it appears that there is sufficient evidence to warrant rejection of the claim that when the 13th day of a month falls on a Friday, the numbers of hospital admissions from motor vehicle crashes are not affected. Hospital admissions do appear to be affected.

Summary statistics:

Variable	Observations	Obs. with missing data	Obs. without missing data	Minimum	Maximum	Mean	Std. deviation
6TH	6	0	6	3.0000	11.0000	7.5000	3.3317
13TH	6	0	6	4.0000	14.0000	10.8333	3.6009

t-test for two paired samples / Two-tailed test:

95% confidence interval on the difference between the means:

(-6.4935 , -0.1731)

Difference	-3.3333
t (Observed value)	-2.7116
|t| (Critical value)	2.5708
DF	5
p-value (Two-tailed)	0.0422
alpha	0.05

17. H_0: $\mu_d = 0$. H_1: $\mu_d < 0$. Enter the label **PHOENIX** in cell A1 of an Excel worksheet followed by the amounts grossed data given in the exercise. Enter the label **PRICE** in cell B1 followed by the amounts grossed data given in the exercise. Select the **XLSTAT** add-in. Select **Parametric tests**. Select **Two-sample *t*-test and *z*-test**. Enter the range of the Phoenix data, including the label (**A1:A11**), in the sample 1 window. Enter the range of the Prince data, including the label (**B1:B11**), in the sample 2 window. Select **Paired samples**. Select **Column labels**. Select **Student's *t* test**. Click the **Options** tab. Alternative hypothesis: **Mean 1 - Mean 2 < D**.

Hypothesized difference (D): **0**. Significance level (%): **5**. Click **OK**. The summary statistics are displayed below. The test results are displayed below. Test statistic: t (Observed value) = -1.0803. t (Critical value) = -1.8331. P-value (one-tailed) = 0.1540. Fail to reject H_0. There is not sufficient evidence to support the claim that *Harry Potter and the Half-Blood Prince* did better at the box office. After a few years, the gross amounts from both movies can be identified, and the conclusion can then be judged objectively without using a hypothesis test.

Summary statistics:

Variable	Observations	Obs. with missing data	Obs. without missing data	Minimum	Maximum	Mean	Std. deviation
PHOENIX	10	0	10	7.6000	44.2000	18.5400	11.8965
PRINCE	10	0	10	6.9000	58.2000	20.1100	15.8179

t-test for two paired samples / Lower-tailed test:

95% confidence interval on the difference between the means:

(-Inf , 1.0940)	

Difference	-1.5700
t (Observed value)	-1.0803
t (Critical value)	-1.8331
DF	9
p-value (one-tailed)	0.1540
alpha	0.05

19. Enter the label **BEFORE** in cell A1 of an Excel worksheet followed by the before pain scale data given in the exercise. Enter the label **AFTER** in cell B1 followed by the after pain scale data given in the exercise. Select the **XLSTAT** add-in. Select **Parametric tests**. Select **Two-sample *t*-test and *z*-test**. Enter the range of the before data, including the label (**A1:A9**), in the sample 1 window. Enter the range of the after data, including the label (**B1:B9**) in the sample 2 window. Select **Paired samples**. Select **Column labels**. Select **Student's *t* test**. Click the **Options** tab. Alternative hypothesis: **Mean 1 - Mean 2 $\neq$ D**. Hypothesized difference (D): **0**. Significance level (%): **5**. Click **OK**. The output is displayed below. The 95% CI: $0.69 < \mu_d < 5.56$. Because the confidence interval limits do not contain 0 and they consist of positive values only, it appears that the "before" measurements are greater than the "after" measurements, so hypnotism does appear to be effective in reducing pain.

Summary statistics:

Variable	Observations	Obs. with missing data	Obs. without missing data	Minimum	Maximum	Mean	Std. deviation
BEFORE	8	0	8	6.3000	11.6000	8.7125	2.1774
AFTER	8	0	8	2.0000	8.5000	5.5875	2.6085

t-test for two paired samples / Two-tailed test:

95% confidence interval on the difference between the means:

(0.6893, 5.5607)	

Difference	3.1250
t (Observed value)	3.0359
|t| (Critical value)	2.3662
DF	7
p-value (Two-tailed)	0.0190
alpha	0.05

21. H_0: $\mu_d = 0$. H_1: $\mu_d \neq 0$. Open the **OSCR** data file on your data disk. Select the **XLSTAT** add-in. Select **Parametric tests**. Select **Two-sample t-test and z-test**. Enter the range of the actresses' data, including the label (**A1:A83**), in the sample 1 window. Enter the range of the actors' data, including the label (**B1:B83**), in the sample 2 window. Select **Paired samples**. Select **Column labels**. Select **Student's t test**. Click the **Options** tab. Alternative hypothesis: **Mean 1 - Mean 2 $\neq$ D**. Hypothesized difference (D): **0**. Significance level (%): **5**. Click **OK**. The output is displayed below. Test statistic: t (Observed value) = -5.5533. t (Critical values) = ± 1.9897. P-value < 0.0001. Reject H_0. There is sufficient evidence to support the claim that there is a difference between the ages of actresses and actors when they win Oscars.

Summary statistics:

Variable	Observations	Obs. with missing data	Obs. without missing data	Minimum	Maximum	Mean	Std. deviation
Actresses	82	0	82	21.0000	80.0000	35.8780	11.1127
Actors	82	0	82	29.0000	76.0000	44.1220	8.9957

t-test for two paired samples / Two-tailed test:

95% confidence interval on the difference between the means:
(-11.1976 , -5.2902)

Difference	-8.2439
t (Observed value)	-5.5533
|t| (Critical value)	1.9897
DF	81
p-value (Two-tailed)	< 0.0001
alpha	0.05

23. Assume you are testing that the males speak fewer words in a day than females. H_0: $\mu_d = 0$. H_1: $\mu_d < 0$. Open the **WORDS** data file on your data disk. Select the **XLSTAT** add-in. Select **Parametric tests**. Select **Two-sample t-test and z-test**. Enter the range of the M1 data, including the label (**A1:A57**), in the sample 1 window. Enter the range of the F1 data, including the label (**B1:B57**), in the sample 2 window. Select **Paired samples**. Select **Column labels**. Select **Student's t test**. Click the **Options** tab. Alternative hypothesis: **Mean 1 - Mean 2 < D**. Hypothesized difference (D): **0**. Significance level (%): **5**. Click **OK**. Test statistic: t (Observed value) = -1.5602. t (Critical value) = -1.6730. P-value (one-tailed) = 0.0622. Fail to reject H_0. There is not sufficient evidence to support the claim that among couples, males speak fewer words in a day than females.

Summary statistics:

Variable	Observations	Obs. with missing	Obs. without missing data	Minimum	Maximum	Mean	Std. deviation
M1	56	0	56	1411.2329	46978.2857	16576.1060	7871.4902
F1	56	0	56	5198.2569	38153.6524	18443.1986	7459.6103

t-test for two paired samples / Lower-tailed test:

95% confidence interval on the difference between the means:
(-Inf , 134.9957)

Difference	-1867.0926
t (Observed value)	-1.5602
t (Critical value)	-1.6730
DF	55
p-value (one-tailed)	0.0622
alpha	0.05

Section 9-4, Beyond the Basics

25. H_0: $\mu_d = 6.8\,\text{kg}$. H_1: $\mu_d \neq 6.8\,\text{kg}$. Select the **XLSTAT** add-in. Select **Parametric tests**. Select **Two-sample t-test and z-test**. Enter the range of the WTAPR data, including the label (**C1:C68**), in the sample 1 window. Enter the range of the WTSEP data, including the label (**B1:B68**), in the sample 2 window. Select **Paired samples**. Select **Column labels**. Select **Student's t test**. Click the **Options** tab. Alternative hypothesis: **Mean 1 - Mean 2 $\neq$ D**. Hypothesized difference (D): **6.8**. Significance level (%): **5**. Click **OK**. The output is displayed below. Test statistic: t (Observed value) = -11.8327. t (Critical values) = ±1.9966. P-value (Two-tailed) < 0.0001 Reject H_0. There is sufficient evidence to warrant rejection of the claim that $\mu_d = 6.8\,\text{kg}$. It appears that the "Freshman 15" is a myth, and college freshman might gain some weight, but they do not gain as much as 15 pounds.

Summary statistics:

Variable	Observations	Obs. with missing data	Obs. without missing data	Minimum	Maximum	Mean	Std. deviation
WTAPR	67	0	67	47.0000	105.0000	66.2388	11.2843
WTSEP	67	0	67	42.0000	97.0000	65.0597	11.2854

t-test for two paired samples / Two-tailed test:

95% confidence interval on the difference between the means:
(0.2307, 2.1275)

Difference	1.1791
t (Observed value)	-11.8327
|t| (Critical value)	1.9966
DF	66
p-value (Two-tailed)	< 0.0001
alpha	0.05

Section 9-5, Basic Skills and Concepts

1. a. No.   b. No.

   c. The two samples have the same standard deviation (or variance).

3. The F test is very sensitive to departures from normality, which means that it works poorly by leading to wrong conclusions when either or both of the populations has a distribution that is not normal. The F test is not robust against sampling methods that do not produce simple random samples. For example, conclusions based on voluntary response samples could easily be wrong.

5. H_0: $\sigma_1 = \sigma_2$. H_1: $\sigma_1 \neq \sigma_2$. Select the **XLSTAT** add-in. Select **Parametric tests**. Select **Two-sample comparison of variances**. Enter the range of the CKWEGWT data, including the label (**A1:A37**), in the sample 1 window. Enter the range of the PPWEGWT data, including the label (**E1:E37**), in the sample 2 window. Select **One column per sample**. Select **Column labels**. Select **Fisher's F-test**. Click the **Options** tab. Alternative hypothesis: **Variance 1/Variance 2 $\neq$ R**. Hypothesized ratio (R): **1**. Significance level (%): **5**. Click **OK**. Test statistic: F (Observed value) = 1.7341. Upper critical F value = 1.9611. P-value = 0.1081. Fail to reject H_0. There is not sufficient evidence to support the claim that weights of regular Coke and weights of regular Pepsi have different standard deviations.

Summary statistics:

Variable	Observations	Obs. with missing data	Obs. without missing data	Minimum	Maximum	Mean	Std. deviation
CKREGWT	36	0	36	0.7901	0.8295	0.8168	0.0075
PPREGWT	36	0	36	0.8139	0.8401	0.8241	0.0057

Fisher's F-test / Two-tailed test:			
95% confidence interval on the ratio of variances:			
(0.8842, 3.4007)			
Ratio	1.7341		
F (Observed value)	1.7341		
F (Critical value)	1.9611		
DF1	35		
DF2	35		
p-value (Two-tailed)	0.1081		
alpha	0.05		

7. H_0: $\sigma_1 = \sigma_2$. H_1: $\sigma_1 \neq \sigma_2$. Test statistic: $F = \dfrac{5.90^2}{5.48^2} = 1.1592$. Using Excel's F.INV.RT function, the upper

critical F value =**F.INV.RT(0.025,34,35)** = 1.9611. Using Excel's F.DIST.RT function, the P-value
=**2*F.DIST.RT(1.1592,34,35)** = 0.6655. Fail to reject H_0. There is not sufficient evidence to warrant rejection of the claim that the samples are from populations with the same standard deviation. The background color does not appear to have an effect on the variation of word recall scores.

9. H_0: $\sigma_1 = \sigma_2$. H_1: $\sigma_1 > \sigma_2$. Test statistic: $F = \dfrac{2.2^2}{0.72^2} = 9.3364$. Using Excel's F.INV.RT function, the critical F

value =**(F.INV.RT(0.05,21,21)** = 2.0842. Using Excel's F.DIST.RT function, the P-value
=**F.DIST.RT(9.3364,21,21)** = 1.67929E-06. Reject H_0. There is sufficient evidence to support the claim that the treatment group has errors that vary more than the errors of the placebo group.

11. H_0: $\sigma_1 = \sigma_2$. H_1: $\sigma_1 > \sigma_2$. Test statistic: $F = \dfrac{1.4^2}{0.96^2} = 2.1267$. Using Excel's F.INV.RT function, the critical F

value =**(F.INV.RT(0.05,19,19)** = 2.1683. Using Excel's F.DIST.RT function, the P-value
=**F.DIST.RT(2.1267,19,19)** = 0.0543. Fail to reject H_0. There is not sufficient evidence to support the claim that those given a sham treatment (similar to a placebo) have pain reductions that vary more than the pain reductions for those treated with magnets.

13. H_0: $\sigma_1 = \sigma_2$. H_1: $\sigma_1 > \sigma_2$. Enter the label **Pennsylvania** in cell A1 of an Excel worksheet followed by the Pennsylvania strontium-90 data. Enter the label **New York** in cell B1 followed by the New York strontium-90 data. Select the **XLSTAT** add-in. Select **Parametric tests**. Select **Two-sample comparison of variances**. Enter the range of the Pennsylvania data, including the label, in the sample 1 window. Enter the range of the New York data, including the label, in the sample 2 window. Select **One column per sample**. Select **Column labels**. Select **Fisher's F-test**. Click the **Options** tab. Alternative hypothesis: **Variance 1/Variance 2 > R**. Hypothesized ratio (R): **1**. Significance level (%): **5**. Click **OK**. Test statistic: F (Observed value) = 4.1648. F (Critical Value) = 2.8179. P-value (one-tailed) = 0.0130. Reject H_0. There is sufficient evidence to support the claim that amounts of strontium-90 from Pennsylvania residents vary more than amounts from New York residents.

Summary statistics:							
Variable	Observations	Obs. with missing data	Obs. without missing data	Minimum	Maximum	Mean	Std. deviation
Pennsylvania	12	0	12	130.0000	163.0000	147.5833	10.6383
New York	12	0	12	128.0000	143.0000	136.4167	5.2129

Fisher's F-test / Upper-tailed test:			
95% confidence interval on the ratio of variances:			
(11.7360, +Inf)			
Ratio	4.1648		
F (Observed value)	4.1648		
F (Critical value)	2.8179		
DF1	11		
DF2	11		
p-value (one-tailed)	0.0130		
alpha	0.05		

15. H_0: $\sigma_1 = \sigma_2$. H_1: $\sigma_1 \neq \sigma_2$. Enter the label **Female** in cell A1 of an Excel worksheet followed by the females' height data. Enter the label **Male** in cell B1 followed by the males' height data. Select the **XLSTAT** add-in. Select **Parametric tests**. Select **Two-sample comparison of variances**. *(Note that, for a two-tailed test, the sample with the larger variance should be entered as sample 1.)* Enter the range of the males' height data, including the label (B1:B11), in the sample 1 window. Enter the range of the females' height data, including the label (A1:A11), in the sample 2 window. Select **One column per sample**. Select **Column labels**. Select **Fisher's F-test**. Click the **Options** tab. Alternative hypothesis: **Variance 1/Variance 2 $\neq$ R**. Hypothesized ratio (R): **1**. Significance level (%): **5**. Click **OK**. Test statistic: F (Observed value) = 1.0073. Upper F (Critical Value) = 4.0260. P-value (Two-tailed) = 0.9915. Fail to reject H_0. There is not sufficient evidence to warrant rejection of the claim that females and males have heights with the same amount of variation.

Summary statistics:

Variable	Observations	Obs. with missing data	Obs. without missing data	Minimum	Maximum	Mean	Std. deviation
Male	10	0	10	169.0000	187.8000	179.7000	6.0647
Female	10	0	10	153.0000	170.9000	161.2300	6.0426

Fisher's F-test / Two-tailed test:

95% confidence interval on the ratio of variances:
(0.2502, 4.0554)

Ratio	1.0073
F (Observed value)	1.0073
F (Critical value)	4.0260
DF1	9
DF2	9
p-value (Two-tailed)	0.9915
alpha	0.05

17. H_0: $\sigma_1 = \sigma_2$. H_1: $\sigma_1 > \sigma_2$. You are testing the claim that females have weights with more variation than males. The females' weight data are in the **FBODY** data file on the data disk, and the males' weight data are in the **MBODY** data file. Select the **XLSTAT** add-in. Select **Parametric tests**. Select **Two-sample comparison of variances**. Enter the range of the males' weight data, including the label, in the sample 1 window. Enter the range of the females' weight data, including the label, in the sample 2 window. Select **One column per sample**. Select **Column labels**. Select **Fisher's F-test**. Click the **Options** tab. Alternative hypothesis: **Variance 1/Variance 2 > R**. Hypothesized ratio (R): **1**. Significance level (%): **5**. Click **OK**. Test statistic: F (Observed value) = 1.2468. F (Critical Value) = 1.7045. P-value (one-tailed) = 0.2471. Fail to reject H_0. There is not sufficient evidence to support the claim that females have weights with more variation than males.

Summary statistics:

Variable	Observations	Obs. with missing data	Obs. without missing data	Minimum	Maximum	Mean	Std. deviation
FEMALE-WT	40	0	40	43.8000	126.6000	74.8275	20.6883
MALE-WT	40	0	40	56.2000	134.3000	82.9750	18.5281

Fisher's F-test / Upper-tailed test:

95% confidence interval on the ratio of variances:
(2.1251, +Inf)

Ratio	1.2468
F (Observed value)	1.2468
F (Critical value)	1.7045
DF1	39
DF2	39
p-value (one-tailed)	0.2471
alpha	0.05

Section 9-5, Beyond the Basics

19. a. No solution provided.

    b. $c_1 = 4$, $c_2 = 0$

    c. Critical value $= \dfrac{\log(0.05/2)}{\log\left(\dfrac{40}{40+40}\right)} = 5$.

    d. Fail to reject $\sigma_1^2 = \sigma_2^2$.

21. $F_L = 0.2727$, $F_R = 2.8365$

Chapter Quick Quiz

1. H_0: $p_1 = p_2$. H_1: $p_1 \neq p_2$.

2. $\overline{p} = \dfrac{347+305}{386+359} = 0.875$

3. Using Excel's NORM.S.DIST function, the one-tailed P-value $= P(z > 2.04)$ **=1-NORM.S.DIST(2.04,TRUE)** $= 0.0207$. The two-tailed P-value $= 2 \cdot P(z > 2.04) = 0.0414$.

4. $0.00172 < p_1 - p_2 < 0.0970$

5. Because the data consist of matched pairs, they are dependent.

6. H_0: $\mu_d = 0$. H_1: $\mu_d > 0$.

7. There is not sufficient evidence to support the claim that front repair costs are greater than the corresponding rear repair costs.

8. F distribution

9. False.

10. True.

Review Exercises

1. H_0: $p_1 = p_2$. H_1: $p_1 > p_2$. Test statistic: $z = 3.12$. Using Excel's NORM.S.INV function, the critical value is **=1-NORM.S.INV(0.99)** $= 2.3263$. Using Excel's NORM.S.DIST function, the P-value **=1-NORM.S.DIST(6.44,TRUE)** $= 0.0009$. Reject H_0. There is sufficient evidence to support a claim that the proportion of successes with surgery is greater than the proportion of successes with splinting. When treating carpal tunnel syndrome, surgery should generally be recommended instead of splinting.

2. 98% CI: $0.0583 < p_1 - p_2 < 0.331$. The confidence interval limits do not contain 0; the interval consists of positive values only. This suggests that the success rate with surgery is greater than the success rate with splints.

3. H_0: $p_1 = p_2$. H_1: $p_1 < p_2$. Test statistic: $z = -1.91$. Using Excel's NORM.S.INV function, the critical value is **=NORM.S.INV(0.05)** = -1.645. Using Excel's NORM.S.DIST function, the P-value **=NORM.S.DIST(-1.91,TRUE)** $= 0.0281$. Reject H_0. There is sufficient evidence to support the claim that the fatality rate of occupants is lower for those in cars equipped with airbags.

4. H_0: $\mu_d = 0$. H_1: $\mu_d > 0$. Enter the label **1 Day in Advance** in cell A1 followed by 1-day-in-advance flight costs of the seven airlines. Enter the label **30 Days in Advance** in cell B1 followed by the 30-days-in-advance flight costs. Select the **XLSTAT** add-in. Select **Parametric tests**. Select **Two-sample *t*-test and *z*-test**. Enter the range of the 1-day-in-advance costs, including the label (**A1:A8**), in the sample 1 window. Enter the range of the 30-days-in-advance costs, including the label (**B1:B8**), in the sample 2 window. Select **Paired samples**. Select **Column labels**. Select **Student's *t* test**. Click the **Options** tab. Alternative hypothesis: **Mean 1 - Mean 2 > D**. Hypothesized difference (D): **0**. Significance level (%): **1**. Click **OK**. Test statistic: *t* (Observed value) = 4.7118. *t* (Critical value) = 3.1428. *P*-value (one-tailed) = 0.0016. Reject H_0. There is sufficient evidence to

support the claim that flights scheduled 1 day in advance cost more than flights scheduled 30 days in advance. Save money by scheduling flights 30 days in advance.

t-test for two paired samples / Upper-tailed test:		
99% confidence interval on the difference between the means:		
(139.0897, +Inf)		
Difference	417.7143	
t (Observed value)	4.7118	
t (Critical value)	3.1428	
DF	6	
p-value (one-tailed)	0.0016	
alpha	0.01	

5. H_0: $\mu_d = 0$. H_1: $\mu_d \neq 0$. Enter the label **Reported** in cell A1 followed by the reported height data. Enter the label **measured** in cell B1 followed by the measured height data. Select the **XLSTAT** add-in. Select **Parametric tests**. Select **Two-sample _t_-test and _z_-test**. Enter the range of the reported height data, including the label (**A1:A9**), in the sample 1 window. Enter the range of the measured height data, including the label (**B1:B9**) in the sample 2 window. Select **Paired samples**. Select **Column labels**. Select **Student's _t_ test**. Click the **Options** tab. Alternative hypothesis: **Mean 1 - Mean 2 $\neq$ D**. Hypothesized difference (D): **0**. Significance level (%): **5**. Click **OK**. The test results are displayed below. Test statistic: t (Observed value) = -0.5739. t (Critical value) = ±2.3662. P-value (Two-tailed) = 0.5840. Fail to reject H_0. There is not sufficient evidence to support the claim that there is a difference between self-reported heights and measured heights of females aged 12–16.

t-test for two paired samples / Two-tailed test:		
95% confidence interval on the difference between the means:		
(-2.1774 , 1.3274)		
Difference	-0.4250	
t (Observed value)	-0.5739	
|t| (Critical value)	2.3662	
DF	7	
p-value (Two-tailed)	0.5840	
alpha	0.05	

6. H_0: $\mu_1 = \mu_2$. H_1: $\mu_1 > \mu_2$. Test statistic: $t = 2.879$. Using Excel's T.INV function, the critical value =**T.INV(0.99,78)** = 2.3751. Using Excel's T.DIST.RT function, the P-value =**T.DIST.RT(2.879,78)** = 0.0026. Reject H_0. There is sufficient evidence to support the claim that "stress decreases the amount recalled."

7. Form a 98% CI. Using Excel's T.INV.2T function, critical t =**T.INV.2T(0.02,78)** = 2.3751.

$$E = t_{\alpha/2}\sqrt{\frac{s_1^2}{n_1} + \frac{s_2^2}{n_2}} = 2.3751\sqrt{\frac{11.6^2}{40} + \frac{13.2^2}{40}} = 6.5992$$

98% CI: $(\bar{x}_1 - \bar{x}_2) - E < \mu_1 - \mu_2 < (\bar{x}_1 - \bar{x}_2) + E = (53.3 - 45.3) - 6.5992 < \mu_1 - \mu_2 < (53.3 - 45.3) + 6.5992$

98% CI: $1.4 < \mu_1 - \mu_2 < 14.6$. The confidence interval limits do not contain 0; the interval consists of positive values only. This suggests that the numbers of details recalled are lower for those in the stress population.

8. H_0: $p_1 = p_2$. H_1: $p_1 \neq p_2$. Test statistic: $z = -4.20$. Using Excel's NORM.S.INV function, the critical values =**NORM.S.INV(0.995)** = ±2.5758. Using Excel's NORM.S.DIST function, the P-value =**2*NORM.S.DIST(-4.2,TRUE)** = 2.66915E-05. Reject H_0. There is sufficient evidence to warrant rejection of the claim that the acceptance rate is the same with or without blinding. Without blinding, reviewers know the names and institutions of the abstract authors, and they might be influenced by that knowledge.

9. H_0: $\mu_1 = \mu_2$. H_1: $\mu_1 \neq \mu_2$. Test statistic: $t = 0.679$. Using Excel's T.INV.2T function, the critical values =**T.INV.2T(0.05, 47)** = ± 2.0117. Using Excel's T.DIST.2T function, the P-value =**T.DIST.2T(0.679,47** = 0.5005. Fail to reject H_0. There is not sufficient evidence to warrant rejection of the claim of no difference between the mean LDL cholesterol levels of subjects treated with raw garlic and subjects given placebos. Both groups appear to be about the same.

10. H_0: $\sigma_1 = \sigma_2$. H_1: $\sigma_1 \neq \sigma_2$. Test statistic: $F = \dfrac{15^2}{14^2} = 1.1480$. Using Excel's F.INV.RT function, the upper critical F value =**F.INV.RT(0.025,48,47)** = **1.7799**. Using Excel's F.DIST.RT function, the P-value =**2*F.DIST.RT(1.1480,48,47)** = 0.6371. Fail to reject H_0. There is not sufficient evidence to warrant rejection of the claim that the two populations have LDL levels with the same standard deviation.

Cumulative Review Exercises

1. a. Because the sample data are matched with each column consisting of heights from the same family, the data are dependent.

    b. Enter the label **Daughter** in cell A1 of an Excel worksheet followed by the daughter height data. Click **DATA** at the top of the screen. Select **Data Analysis**. Select **Descriptive Statistics**. Enter the range of the daughters' height data, including the label (**A1:A11**), in the input range window. Select **Labels in first row**. Select **Summary statistics**. Click **OK**. Mean: 63.81 in.; median: 63.70 in.; mode: 62.2 in.; range: 8.80 in.; standard deviation: 2.73 in.; variance: 7.43 in^2

Daughter	
Mean	63.81
Standard Error	0.862097
Median	63.7
Mode	62.2
Standard Deviation	2.72619
Sample Variance	7.432111
Kurtosis	-0.49504
Skewness	0.241088
Range	8.8
Minimum	59.6
Maximum	68.4
Sum	638.1
Count	10

    c. Ratio

2. Enter the mothers' heights and the daughters' heights in an Excel worksheet. Click and drag over the height data to select the data for the scatterplot. Click **INSERT** at the top of the screen. In the Charts group, select **Scatter**. Select the leftmost figure in the top row. To add titles, click **DESIGN** at the top of the screen. In the Chart Layouts group, select **Add Chart Element**. Select **Axis Titles**. Select **Primary Horizontal**. Select **Primary Vertical**. For the primary horizontal, type **Heights of Mothers (in.)**. For the primary vertical, type **Heights of Daughters (in.)**. The scatterplot is displayed on the next page. There does not appear to be a correlation or association between the heights of mothers and the heights of their daughters.

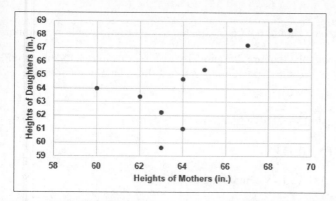

3. Enter the daughters' heights in an Excel worksheet. Click **DATA** at the top of the screen. Select **Data Analysis**. Select **Descriptive Statistics**. Click **OK**. Enter the range of the daughters' height data in the input range window. If you included a label, click **Labels in First Row**. Select **Summary statistics**. Select **Confidence Level for Mean**, and enter **95** for a 95% confidence interval. Click **OK**. The confidence level (95.0%) equals 1.9502. The 95% confidence interval limits are obtained by subtracting and adding 1.9502 to the sample mean (63.81). 95% CI: $(63.81 - 1.9502) < \mu < (63.81 + 1.9502) = 61.86 < \mu < 65.76$.

4. H_0: $\mu_d = 0$. H_1: $\mu_d \neq 0$. Enter the mothers' heights and the daughters' heights in an Excel worksheet. Select the **XLSTAT** add-in. Select **Parametric tests**. Select **Two-sample *t*-test and *z*-test**. Enter the range of the mothers' height data, including the label, in the sample 1 window. Enter the range of the daughters' height data, including the label, in the sample 2 window. Select **Paired samples**. Select **Column labels**. Select **Student's *t* test**. Click the **Options** tab. Alternative hypothesis: **Mean 1 - Mean 2 ≠ D**. Hypothesized difference (D): **0**. Significance level (%): **5**. Click **OK**. Test statistic: *t* (Observed value) = 0.2833. *t* (Critical value) = ±2.2622. *P*-value (Two-tailed) = 0.7834. Fail to reject H_0. There is not sufficient evidence to warrant rejection of the claim of no significant difference between the heights of mothers and the heights of their daughters.

Summary statistics:

Variable	Observations	Obs. with missing data	Obs. without missing data	Minimum	Maximum	Mean	Std. deviation
Mother	10	0	10	60.0000	69.0000	64.0000	2.5386
Daughter	10	0	10	59.6000	68.4000	63.8100	2.7262

t-test for two paired samples / Two-tailed test:

95% confidence interval on the difference between the means:

(-1.3273 , 1.7073)

Difference	0.1900
t (Observed value)	0.2833
|t| (Critical value)	2.2622
DF	9
p-value (Two-tailed)	0.7834
alpha	0.05

5. Because the points lie reasonably close to a straight-line pattern and there is no other pattern that is not a straight-line pattern and there are no outliers, the sample data appear to be from a population with a normal distribution.

6. $0.109 < p_1 < 0.150$. Because the entire range of values in the confidence interval lies below 0.20, the results do justify the statement that "fewer than 20% of Americans choose their computer and/or Internet access when identifying what they miss most when electrical power is lost."

7. No. Because the Internet users chose to respond, we have a voluntary response sample, so the results are not necessarily valid.

8. $n = \dfrac{[z_{\alpha/2}]^2 \, \hat{p}\hat{q}}{E^2} = \dfrac{[2.17]^2 \, (0.25)}{0.02^2} = 2944$. The survey should not be conducted using only local phone numbers.

Such a convenience sample could easily lead to results that are dramatically different from results that would be obtained by randomly selecting respondents from the entire population, not just those having local phone numbers.

9.   a. $z = \dfrac{152.1 - 162.0}{6.6} = -1.5$. Using Excel's NORM.S.DIST function,

$P(z > -1.5)$ **=1-NORM.S.DIST(-1.5,TRUE)** $= 0.9332$.

b. $z = \dfrac{152.1 - 162.0}{6.6/\sqrt{4}} = -3$. Using Excel's NORM.S.DIST function,

$P(z > -3)$ **=1-NORM.S.DIST(-3,TRUE)** $= 0.9987$

c.   Using Excel's NORM.INV function, the 80th percentile **=NORM.INV(0.8,162.0,6.6)** = 167.6 cm.

10. No. Because the states have different population sizes, the mean cannot be found by adding the 50 state means and dividing the total by 50. The mean income for the U.S. population can be found by using a weighted mean that incorporates the population size of each state.

Chapter 10

Correlation and Regression

Section 10-2, Basic Skills and Concepts

1. r represents the value of the linear correlation computed by using the paired sample data. ρ represents the value of the linear correlation coefficient that would be computed by using all of the paired data in the population. The value of r is estimated to be 0 (because there is no correlation between sunspot numbers and the Dow Jones Industrial Average).

3. The headline is not justified because it states that increased salt consumption is the cause of higher blood pressure levels, but the presence of a correlation between two variables does not necessarily imply that one is the cause of the other. Correlation does not imply causality. A correct headline would be this: "Study Shows That Increased Salt Consumption Is Associated with Higher Blood Pressure."

5. H_0: $\rho = 0$. H_1: $\rho \neq 0$; Yes. With $r = 0.687$ and critical values of ± 0.312, there is sufficient evidence to support the claim that there is a linear correlation between the durations of eruptions and the time intervals to the next eruptions.

7. H_0: $\rho = 0$. H_1: $\rho \neq 0$; No. With r = 0.149 and a P-value of 0.681 (or critical values of ± 0.632), there is not sufficient evidence to support the claim that there is a linear correlation between the heights of fathers and the heights of their sons.

9. a. Enter the x and y data in an Excel worksheet. Click and drag over the cells that contain the data to select them for the chart. Click **INSERT** on the Ribbon. Click **Scatter** in the Charts section. Select the leftmost figure in the top row.

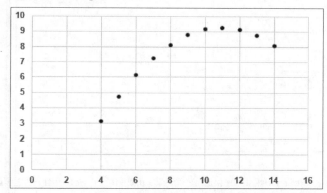

   b. To use Excel's CORREL function to find r, click **FORMULAS** on the Ribbon. Select **Insert Function**. Select **Statistical**. Select **CORREL**. Click **OK**. Click on the **Array 1** box and enter the range of the x data. Click on the Array 2 box and enter the range of the y data. Click **OK**. $r = 0.816$. H_0: $\rho = 0$. H_1: $\rho \neq 0$. Critical values: $r = \pm 0.602$. There is sufficient evidence to support the claim of a linear correlation between the two variables.

   c. The scatterplot reveals a distinct pattern that is not a straight line pattern.

11. a. There appears to be a linear correlation.

   b. Enter the x and y data in an Excel worksheet. To use Excel's CORREL function to find r, click **FORMULAS** on the Ribbon. Select **Insert Function**. Select **Statistical**. Select **CORREL**. Click **OK**. Click on the **Array 1** box and enter the range of the x data. Click on the **Array 2** box and enter the range of the y data. Click **OK**. $r = 0.906$. H_0: $\rho = 0$. H_1: $\rho \neq 0$. Critical values: $r = \pm 0.632$ (for a 0.05 significance level). There is a linear correlation.

   c. To use Excel's CORREL function to find r, click **FORMULAS** on the Ribbon. Select **Insert Function**. Select **Statistical**. Select **CORREL**. Click **OK**. Click on the **Array 1** box and enter the range of the x data not including the point with coordinates (10,10). Click on the **Array 2** box and enter the range of the y data not including the point with coordinates (10,10). Click **OK**. $r = 0$. H_0: $\rho = 0$. H_1: $\rho \neq 0$. Critical values: $r = \pm 0.666$ (for a 0.05 significance level). There does not appear to be a linear correlation.

13. Enter the lemon imports and crash fatality rate data in an Excel worksheet and be sure to use labels. Select the **XLSTAT** add-in. Select **Correlation/Association tests**. Select **Correlation tests**. Enter the data range in the Observations/variables table box. Check the **Variable labels** box. Type of correlation: **Pearson**. Click the **Outputs** tab. Select **Correlations** and **p-values**. Click the **Charts** tab. Select **Scatter plots**. Click **OK**. H_0: $\rho = 0$. H_1: $\rho \neq 0$; $r = -0.959$. Critical values: $r = \pm 0.878$. P-value = 0.0099. There is sufficient evidence to support the claim that there is a linear correlation between weights of lemon imports from Mexico and U.S. car fatality rates. The results do not suggest any cause-effect relationship between the two variables.

p-values:		
Variables	**Lemon Imports**	**Crash Fatality Rate**
Lemon Imports	0	0.0099
Crash Fatality Rate	0.0099	0

Correlation matrix (Pearson):		
Variables	**Lemon Imports**	**Crash Fatality Rate**
Lemon Imports	1	-0.9590
Crash Fatality Rate	-0.9590	1

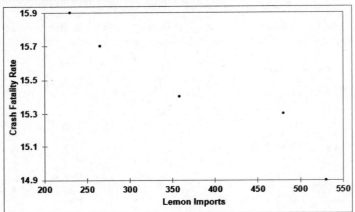

15. Enter the enrollment and burglaries data in an Excel worksheet and be sure to use labels. Select the **XLSTAT** add-in. Select **Correlation/Association tests**. Select **Correlation tests**. Enter the data range in the Observations/variables table box. Check the **Variable labels** box. Type of correlation: **Pearson**. Click the **Outputs** tab. Select **Correlations** and **p-values**. Click the **Charts** tab. Select **Scatter plots**. Click **OK**. H_0: $\rho = 0$. H_1: $\rho \neq 0$; $r = 0.561$. Critical values: $r = \pm 0.632$. P-value = 0.0914. There is not sufficient evidence to support the claim that there is a linear correlation between enrollment and burglaries. The results do not change if the actual enrollments are listed as 32,000, 31,000, 53,000, etc.

Correlation matrix (Pearson):		
Variables	**Enrollment**	**Burglaries**
Enrollment	1	0.5613
Burglaries	0.5613	1

p-values:		
Variables	**Enrollment**	**Burglaries**
Enrollment	0	0.0914
Burglaries	0.0914	0

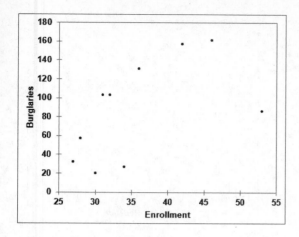

17. Enter the court income and justice salary data in an Excel worksheet and be sure to use labels. Select the **XLSTAT** add-in. Select **Correlation/Association tests**. Select **Correlation tests**. Enter the data range in the Observations/variables table box. Check the **Variable labels** box. Type of correlation: **Pearson**. Click the **Outputs** tab. Select **Correlations** and **p-values**. Click the **Charts** tab. Select **Scatter plots**. Click **OK**. H_0: $\rho = 0$. H_1: $\rho \neq 0$; $r = 0.864$. Critical values: $r = \pm 0.666$. P-value = 0.0026. There is sufficient evidence to support the claim that there is a linear correlation between court incomes and justice salaries. The correlation does not imply that court incomes directly affect justice salaries, but it does appear that justices might profit by levying larger fines, or perhaps justices with higher salaries impose larger fines.

Correlation matrix (Pearson):		
Variables	Court Income	Justice Salary
Court Income	1	0.8643
Justice Salary	0.8643	1

p-values:		
Variables	Court Income	Justice Salary
Court Income	0	0.0026
Justice Salary	0.0026	0

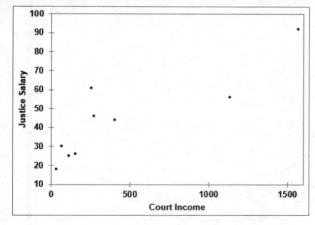

19. Enter the redshift and distance data in an Excel worksheet and be sure to use labels. Select the **XLSTAT** add-in. Select **Correlation/Association tests**. Select **Correlation tests**. Enter the data range in the Observations/variables table box. Check the **Variable labels** box. Type of correlation: **Pearson**. Click the **Outputs** tab. Select **Correlations** and **p-values**. Click the **Charts** tab. Select **Scatter plots**. Click **OK**. H_0: $\rho = 0$. H_1: $\rho \neq 0$; $r = 1.000$. Critical values: $r = \pm 0.811$. P-value < 0.0001. There is sufficient evidence to support the claim that there is a linear correlation between amounts of redshift and distances to clusters of galaxies. Because the linear correlation coefficient is 1.000, it appears that the distances can be directly computed from the amounts of redshift.

Correlation matrix (Pearson):		
Variables	Redshift	Distance
Redshift	1	1.0000
Distance	1.0000	1

p-values:		
Variables	Redshift	Distance
Redshift	0	< 0.0001
Distance	< 0.0001	0

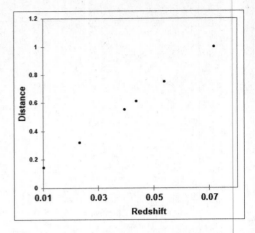

21. Enter the overhead width and weight in an Excel worksheet and be sure to use labels. Select the **XLSTAT** add-in. Select **Correlation/Association tests**. Select **Correlation tests**. Enter the data range in the Observations/variables table box. Check the **Variable labels** box. Type of correlation: **Pearson**. Click the **Outputs** tab. Select **Correlations** and **p-values**. Click the **Charts** tab. Select **Scatter plots**. Click **OK**. H_0: $\rho = 0$. H_1: $\rho \neq 0$; $r = 0.949$. Critical values: r = ±0.811. *P*-value = 0.0039. There is sufficient evidence to support the claim of a linear correlation between the overhead width of a seal in a photograph and the weight of a seal.

Correlation matrix (Pearson):		
Variables	Overhead Width	Weight
Overhead Width	1	0.9485
Weight	0.9485	1

p-values:		
Variables	Overhead Width	Weight
Overhead Width	0	0.0039
Weight	0.0039	0

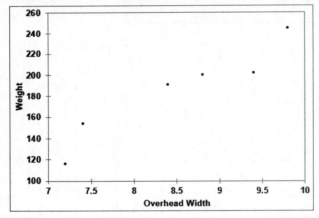

23. Enter the right arm and left arm data in an Excel worksheet and be sure to use labels. Select the **XLSTAT** add-in. Select **Correlation/Association tests**. Select **Correlation tests**. Enter the data range in the Observations/variables table box. Check the **Variable labels** box. Type of correlation: **Pearson**. Click the **Outputs** tab. Select **Correlations** and **p-values**. Click the **Charts** tab. Select **Scatter plots**. Click **OK**. H_0: $\rho = 0$. H_1: $\rho \neq 0$; $r = 0.867$. Critical values: $r = ±0.878$. *P*-value = 0.0570. There is not sufficient evidence to support the claim of a linear correlation between the systolic blood pressure measurements of the right and left arm.

Correlation matrix (Pearson):		
Variables	Right Arm	Left Arm
Right Arm	1	0.8671
Left Arm	0.8671	1

p-values:		
Variables	Right Arm	Left Arm
Right Arm	0	0.0570
Left Arm	0.0570	0

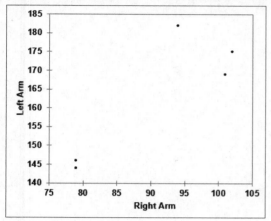

25. Enter the regular and premium data in an Excel worksheet and be sure to use labels. Select the **XLSTAT** add-in. Select **Correlation/Association tests**. Select **Correlation tests**. Enter the data range in the Observations/variables table box. Check the **Variable labels** box. Type of correlation: **Pearson**. Click the **Outputs** tab. Select **Correlations** and **p-values**. Click the **Charts** tab. Select **Scatter plots**. Click **OK**. H_0: $\rho = 0$. H_1: $\rho \neq 0$; $r = 0.197$. Critical values: $r = \pm 0.707$. P-value = 0.6402. There is not sufficient evidence to support the claim that there is a linear correlation between prices of regular gas and prices of premium gas. Because there does not appear to be a linear correlation between prices of regular and premium gas, knowing the price of regular gas is not very helpful in getting a good sense for the price of premium gas.

Correlation matrix (Pearson):		
Variables	Regular	Premium
Regular	1	0.1969
Premium	0.1969	1

p-values:		
Variables	Regular	Premium
Regular	0	0.6402
Premium	0.6402	0

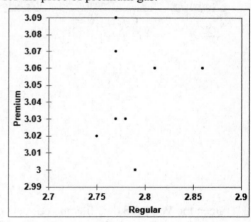

27. Enter the diameter and circumference data in an Excel worksheet and be sure to use labels. Select the **XLSTAT** add-in. Select **Correlation/Association tests**. Select **Correlation tests**. Enter the data range in the Observations/variables table box. Check the **Variable labels** box. Type of correlation: **Pearson**. Click the **Outputs** tab. Select **Correlations** and **p-values**. Click the **Charts** tab. Select **Scatter plots**. Click **OK**. H_0: $\rho = 0$. H_1: $\rho \neq 0$; $r = 1.000$. Critical values: $r = \pm 0.707$. P-value < 0.0001. There is sufficient evidence to support the claim that there is a linear correlation between diameters and circumferences. A scatterplot confirms that there is a linear association between diameters and volumes.

Correlation matrix (Pearson):		
Variables	Diameter	Circumference
Diameter	1	1.0000
Circumference	1.0000	1

p-values:		
Variables	Diameter	Circumference
Diameter	0	< 0.0001
Circumference	< 0.0001	0

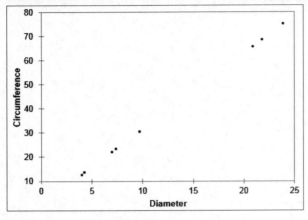

29. Open the **IQBRAIN** data file on your data disk. Select the **XLSTAT** add-in. Select **Correlation/Association tests**. Select **Correlation tests**. Enter the data range in the Observations/variables table box. Check the **Variable labels** box. Type of correlation: **Pearson**. Click the **Outputs** tab. Select **Correlations** and **p-values**. Click the **Charts** tab. Select **Scatter plots**. Click **OK**. H_0: $\rho = 0$. H_1: $\rho \neq 0$; r = -0.063. Critical values: $r = \pm 0.444$. *P*-value = 0.7906. There is not sufficient evidence to support the claim of a linear correlation between IQ and brain volume.

Correlation matrix (Pearson):		
Variables	IQ	VOL
IQ	1	-0.0634
VOL	-0.0634	1

p-values:		
Variables	IQ	VOL
IQ	0	0.7906
VOL	0.7906	0

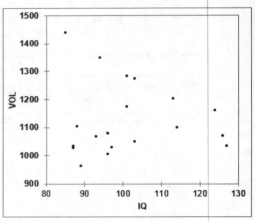

31. Open the **WORDS** data file on your data disk. Select the **XLSTAT** add-in. Select **Correlation/Association tests**. Select **Correlation tests**. Enter the data range in the Observations/variables table box. Check the **Variable labels** box. Type of correlation: **Pearson**. Click the **Outputs** tab. Select **Correlations** and **p-values**. Click the **Charts** tab. Select **Scatter plots**. Click **OK** H_0: $\rho = 0$. H_1: $\rho \neq 0$; $r = 0.319$. Critical values: $r = \pm 0.263$. *P*-value = 0.0167. There is sufficient evidence to support the claim of a linear correlation between the numbers of words spoken by men and women who are in couple relationships.

Correlation matrix (Pearson):		
Variables	M1	F1
M1	1	0.3186
F1	0.3186	1

p-values:		
Variables	M1	F1
M1	0	0.0167
F1	0.0167	0

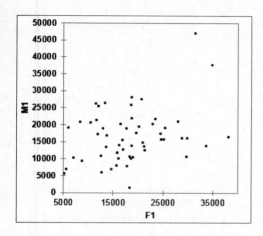

Section 10-2, Beyond the Basics

33. a. Enter the x and y data in an Excel worksheet and be sure to use labels. Select the **XLSTAT** add-in. Select **Correlation/Association tests**. Select **Correlation tests**. Enter the data range in the Observations/variables table box. Check the **Variable labels** box. Type of correlation: **Pearson**. Click the **Outputs** tab. Select **Correlations** and **p-values**. Click the **Charts** tab. Select **Scatter plots**. Click **OK**. H_0: $\rho = 0$. H_1: $\rho \neq 0$; $r = 0.912$. P-value $= 0.0312$. Reject H_0.

Correlation matrix (Pearson):		
Variables	**x**	**y**
x	1	0.9115
y	0.9115	1

p-values:		
Variables	**x**	**y**
x	0	0.0312
y	0.0312	0

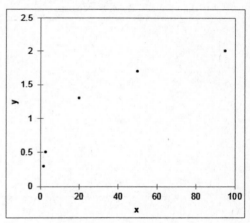

b. Transform the x values to x^2. Select the **XLSTAT** add-in. Select **Correlation/Association tests**. Select **Correlation tests**. Enter the data range in the Observations/variables table box. Check the **Variable labels** box. Type of correlation: **Pearson**. Click the **Outputs** tab. Select **Correlations** and **p-values**. Click the **Charts** tab. Select **Scatter plots**. Click **OK**. H_0: $\rho = 0$. H_1: $\rho \neq 0$; $r = 0.787$. P-value $= 0.1141$. Fail to reject H_0.

Correlation matrix (Pearson):		
Variables	**x^2**	**y**
x^2	1	0.7871
y	0.7871	1

p-values:		
Variables	**x^2**	**y**
x^2	0	0.1141
y	0.1141	0

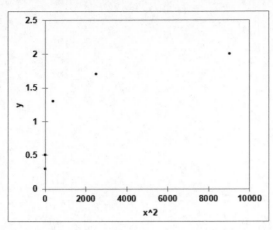

c.   Transform the x values to $\log x$. Select the **XLSTAT** add-in. Select **Correlation/Association tests**. Select **Correlation tests**. Enter the data range in the Observations/variables table box. Check the **Variable labels** box. Type of correlation: **Pearson**. Click the **Outputs** tab. Select **Correlations** and **p-values**. Click the **Charts** tab. Select **Scatter plots**. Click **OK**. H_0: $\rho = 0$. H_1: $\rho \neq 0$; $r = 0.999$. P-value < 0.0001. Reject H_0.

Correlation matrix (Pearson):

Variables	log x	y
log x	1	0.9999
y	0.9999	1

p-values:

Variables	log x	y
log x	0	< 0.0001
y	< 0.0001	0

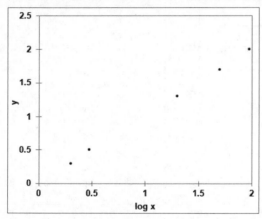

d.   Transform the x values to $\sqrt{x}$. Select the **XLSTAT** add-in. Select **Correlation/Association tests**. Select **Correlation tests**. Enter the data range in the Observations/variables table box. Check the **Variable labels** box. Type of correlation: **Pearson**. Click the **Outputs** tab. Select **Correlations** and **p-values**. Click the **Charts** tab. Select **Scatter plots**. Click **OK**. H_0: $\rho = 0$. H_1: $\rho \neq 0$; $r = 0.976$. P-value $= 0.0045$. Reject H_0.

Correlation matrix (Pearson):

Variables	sqrt(x)	y
sqrt(x)	1	0.9756
y	0.9756	1

p-values:

Variables	sqrt(x)	y
sqrt(x)	0	0.0045
y	0.0045	0

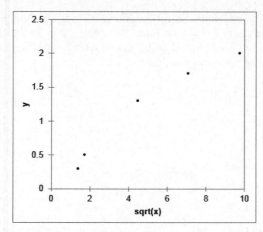

e. Transform the x values to $1/x$. Select the **XLSTAT** add-in. Select **Correlation/Association tests**. Select **Correlation tests**. Enter the data range in the Observations/variables table box. Check the **Variable labels** box. Type of correlation: **Pearson**. Click the **Outputs** tab. Select **Correlations** and **p-values**. Click the **Charts** tab. Select **Scatter plots**. Click **OK**. H_0: $\rho = 0$. H_1: $\rho \neq 0$; $r = -0.948$. P-value = 0.0140. Reject H_0.

Correlation matrix (Pearson):		
Variables	1/x	y
1/x	1	-0.9484
y	-0.9484	1

p-values:		
Variables	1/x	y
1/x	0	0.0140
y	0.0140	0

Case c results in the larger values of r, $r = 0.999$.

Section 10-3, Basic Skills and Concepts

1. The symbol $\hat{y}$ represents the predicted pulse rate. The predictor variable represents height. The response variable represents pulse rate.

3. If r is positive, the regression line has a positive slope and rises from left to right. If r is negative, the slope of the regression line is negative and it falls from left to right.

5. The regression line fits the points well, so the best predicted time for an interval after the eruption is $\hat{y} = 47.4 + 0.180(120) = 69$ min.

7. The regression line does not fit the points well, so the best predicted height is $\overline{y} = 68.0$ in.

9. Enter the x and y data in an Excel worksheet and be sure to use labels. Select the **XLSTAT** add-in. Select **Visualizing data**. Select **Scatter plots**. Enter the x data range in the **X** box. Enter the y data range in the **Y** box. Check **Sheet**. Check **Variable** Labels. Click **OK**. Right-click on a plotted point. Select **Add Trendline**. Click **Display Equation on Chart**. The data have a pattern that is not a straight line.

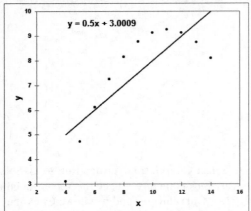

11. a.   Select the **XLSTAT** add-in. Select **Visualizing data**. Select **Scatter plots**. Enter the x data range in the **X** box. Enter the y data range in the **Y** box. Check **Sheet**. Check **Variable** Labels. Click **OK**. Right-click on a plotted point. Select **Add Trendline**. Click **Display Equation on Chart**.

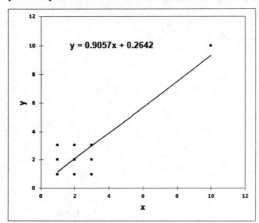

   b.   Repeat the instructions for part a except do not include the point with coordinates (10,10).

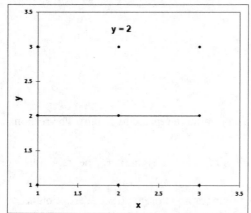

   c.   The results are very different, indicating that one point can dramatically affect the regression equation.

b.

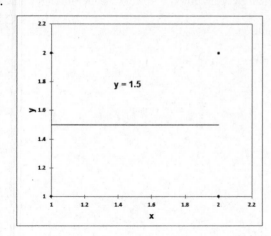

c.

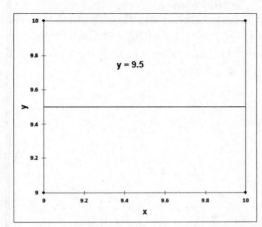

d.   The results are very different, indicating that combinations of clusters can produce results that differ dramatically from results within each cluster alone.

13.   Select the **XLSTAT** add-in. Select **Visualizing data**. Select **Scatter plots**. Enter the *x* data range in the **X** box. Enter the *y* data range in the **Y** box. Check **Sheet**. Check **Variable** Labels. Click **OK**. Right-click on a plotted point. Select **Add Trendline**. Click **Display Equation on Chart**. $\hat{y} = 16.5 - 0.0028x$; The regression line fits the points well, so the best predicted value is $\hat{y} = 16.5 - 0.00282(500) = 15.1$ fatalities per 100,000 population.

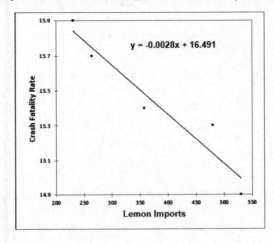

15. Select the **XLSTAT** add-in. Select **Visualizing data**. Select **Scatter plots**. Enter the x data range in the **X** box. Enter the y data range in the **Y** box. Check **Sheet**. Check **Variable** Labels. Click **OK**. Right-click on a plotted point. Select **Add Trendline**. Click **Display Equation on Chart**.
$\hat{y} = -36.8 + 3.47x$; The regression line does not fit the points well, so the best predicted value is $\bar{y} = 87.7$ burglaries. The predicted value is not close to the actual value of 329 burglaries.

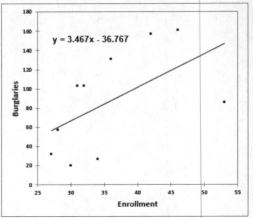

17. Select the **XLSTAT** add-in. Select **Visualizing data**. Select **Scatter plots**. Enter the x data range in the **X** box. Enter the y data range in the **Y** box. Check **Sheet**. Check **Variable** Labels. Click **OK**. Right-click on a plotted point. Select **Add Trendline**. Click **Display Equation on Chart**. $\hat{y} = 27.7 + 0.0373x$; The best predicted value is $\hat{y} = 27.7 + 0.0373(83.941) = 30.8$, which represents \$30,800. The predicted value is not very close to the actual salary of \$26,088. The possible outliers might explain the inaccuracy.

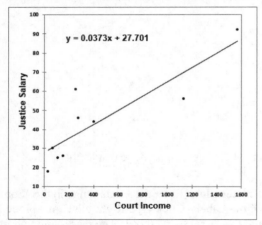

19. Select the **XLSTAT** add-in. Select **Visualizing data**. Select **Scatter plots**. Enter the x data range in the **X** box. Enter the y data range in the **Y** box. Check **Sheet**. Check **Variable** Labels. Click **OK**. Right-click on a plotted point. Select **Add Trendline**.

Click **Display Equation on Chart**. $\hat{y} = -0.00440 + 14.0x$;
The best predicted value is
$\hat{y} = -0.00440 + 14.0(0.0126) = 0.172$ billion light-years.
The predicted value is very close to the actual distance of 0.18 light-years.

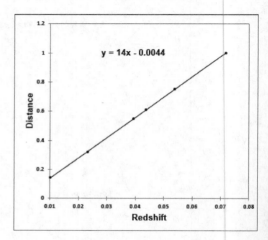

21. Select the **XLSTAT** add-in. Select **Visualizing data**. Select **Scatter plots**. Enter the *x* data range in the **X** box. Enter the *y* data range in the **Y** box. Check **Sheet**. Check **Variable** Labels. Click **OK**. Right-click on a plotted point. Select **Add Trendline**. Click **Display Equation on chart**. Click **Display Equation on Chart**. $\hat{y} = -157 + 40.2x$; The best predicted weight is $\hat{y} = -157 + 40.2(2) = -76.5$ kg. That prediction is a negative weight that cannot be correct. The overhead width of 2 cm is well beyond the scope of the available sample widths, so the extrapolation might be off by a considerable amount.

23. Select the **XLSTAT** add-in. Select **Visualizing data**. Select **Scatter plots**. Enter the *x* data range in the **X** box. Enter the *y* data range in the **Y** box. Check **Sheet**. Check **Variable** Labels. Click **OK**. Right-click on a plotted point. Select **Add Trendline**. Click **Display Equation on chart**. $\hat{y} = 43.6 + 1.31x$; The regression line does not fit the data well, so the best predicted value is $\overline{y} = 163.2$ mm Hg.

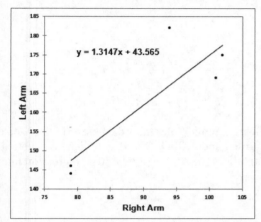

25. Select the **XLSTAT** add-in. Select **Visualizing data**. Select **Scatter plots**. Enter the *x* data range in the **X** box. Enter the *y* data range in the **Y** box. Check **Sheet**. Check **Variable** Labels. Click **OK**. Right-click on a plotted point. Select **Add Trendline**. Click **Display Equation on chart**. $\hat{y} = 2.57 + 0.172x$; The regression line does not fit the data well, so the best predicted value is $\overline{y} = \$3.05$. The predicted price is not very close to the actual price of $2.93.

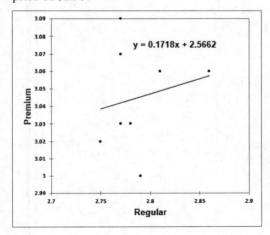

27. Select the **XLSTAT** add-in. Select **Visualizing data**. Select **Scatter plots**. Enter the *x* data range in the **X** box. Enter the *y* data range in the **Y** box. Check **Sheet**. Check **Variable** Labels. Click **OK**. Right-click on a plotted point. Select **Add Trendline**. Click **Display Equation on chart**. $\hat{y} = -0.004 + 3.14x$; The best predicted value is $\hat{y} = -0.004 + 3.14(1.50) = 4.7$ cm. Even though the diameter of 1.50 cm is beyond the scope of the sample diameters, the predicted value yields the actual circumference.

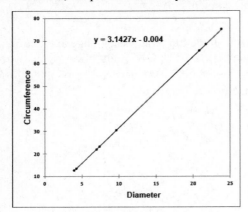

29. Select the **XLSTAT** add-in. Select **Visualizing data**. Select **Scatter plots**. Enter the *x* data range in the **X** box. Enter the *y* data range in the **Y** box. Check **Sheet**. Check **Variable** Labels. Click **OK**. Right-click on a plotted point. Select **Add Trendline**. Click **Display Equation on chart**. $\hat{y} = 108.55 - 0.0067x$; The regression line does not fit the data well, so the best predicted IQ score is $\overline{y} = 101$.

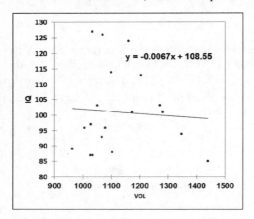

31. Select the **XLSTAT** add-in. Select **Visualizing data**. Select **Scatter plots**. Enter the *x* data range in the **X** box. Enter the *y* data range in the **Y** box. Check **Sheet**. Check **Variable** Labels. Click **OK**. Right-click on a plotted point. Select **Add Trendline**. Click **Display Equation on chart**. $\hat{y} = 13,439 + 0.302x$. The best predicted value is $\hat{y} = 13,439 + 0.302(10,000) = 16,459$.

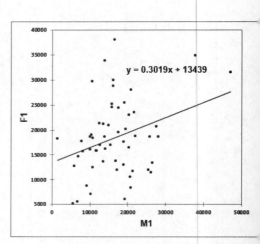

Section 10-3, Beyond the Basics

33. With $\beta_1 = 0$, the regression line is horizontal so that different values of x result in the same y value, and there is no correlation between x and y.

Section 10-4, Basic Skills and Concepts

1. The value of $s_e = 17.5436$ cm is the standard error of estimate, which is a measure of the differences between the observed weights and the weights predicted from the regression equation. It is a measure of the variation of the sample points about the regression line.

3. The coefficient of determination is $r^2 = (0.356)^2 = 0.127$. We know that 12.7% of the variation in weight is explained by the linear correlation between height and weight, and 87.3% of the variation in weight is explained by other factors and/or random variation.

5. $r^2 = (0.933)^2 = 0.870$. 87.0% of the variation in waist size is explained by the linear correlation between weight and waist size, and 13.0% of the variation in waist size is explained by other factors and/or random variation.

7. $r^2 = (-0.793)^2 = 0.629$. 62.9% of the variation in highway fuel consumption is explained by the linear correlation between weight and highway fuel consumption, and 37.1% of the variation in highway fuel consumption is explained by other factors and/or random variation.

9. $r = 0.842$. Critical values: $r = \pm 0.312$ (assuming a 0.05 significance level). *P*-value = 0.000. There is sufficient evidence to support a claim of a linear correlation between foot length and height.

11. $\hat{y} = 64.1 + 4.29(29.0) = 189$ cm

13. Open the **FOOT** data file on your data disk. Select the **XLSTAT** add-in. Select **Modeling data**. Select **Linear regression**. Enter the height data range in the **Y/Dependent variables** box. Enter the foot length data in the **X/Explanatory variables** box. Click **Sheet**. Click **Variable labels**. Click the **Options** tab. Enter **95** in the **Confidence interval(%)** box. Click **OK**. Standard error estimate: RMSE = 5.5057. Using Excel's T.INV.2T function, the *t* critical value =**T.INV.2T(0.05,38)** = 2.0244.

Equation of the model:		
Height = 64.12561+4.29125*Foot Length		

Goodness of fit statistics:	
Observations	40.0000
Sum of weights	40.0000
DF	38.0000
R²	0.7090
Adjusted R²	0.7014
MSE	30.3129
RMSE	5.5057
DW	2.1752

160 cm $< y <$ 183 cm

$$\hat{y} = 64.1 + 4.29(25) = 171.35$$

$$E = t_{a/2}s_e \sqrt{1 + \frac{1}{n} + \frac{n(x_0 - \bar{x})^2}{n(\Sigma x^2) - (\Sigma x)^2}} = 2.024(5.50571)\sqrt{1 + \frac{1}{40} + \frac{40(25 - 25.68)^2}{40(26530.92) - (1027.2)^2}} = 11.299$$

15. Open the **FOOT** data file on your data disk. Select the **XLSTAT** add-in. Select **Modeling data**. Select **Linear regression**. Enter the height data range in the **Y/Dependent variables** box. Enter the foot length data in the **X/Explanatory variables** box. Click **Sheet**. Click **Variable labels**. Click the **Options** tab. Enter **90** in the **Confidence interval(%)** box. Click **OK**. Standard error estimate: RMSE = 5.5057. Using Excel's T.INV.2T function, the *t* critical value =**T.INV.2T(0.10,38)** = 1.6860.

Equation of the model:		
Height = 64.12561+4.29125*Foot Length		

Goodness of fit statistics:

Observations	40.0000
Sum of weights	40.0000
DF	38.0000
R²	0.7090
Adjusted R²	0.7014
MSE	30.3129
RMSE	5.5057
DW	2.1752

$149 \text{ cm} < y < 168 \text{ cm}$

$\hat{y} = 64.1 + 4.29(22) = 158.48$

$$E = t_{a/2}s_e\sqrt{1+\frac{1}{n}+\frac{n(x_0-\bar{x})^2}{n(\Sigma x^2)-(\Sigma x)^2}} = 1.686(5.50571)\sqrt{1+\frac{1}{40}+\frac{40(22-25.68)^2}{40(26530.92)-(1027.2)^2}} = 9.797$$

17. a.   Enter the altitude and temperature data in an Excel worksheet and be sure to include the labels. Select the **XLSTAT** add-in. Select **Modeling data**. Select **Linear regression**. Enter the temperature data range in the **Y/Dependent variables** box. Enter the altitude data in the **X/Explanatory variables** box. Click **Sheet**. Click **Variables labels**. Click the **Outputs** tab. Click **Descriptive statistics** and **Analysis of variance**. Click **OK**.

Goodness of fit statistics:

Observations	7.0000
Sum of weights	7.0000
DF	5.0000
R²	0.9936
Adjusted R²	0.9923
MSE	13.7672
RMSE	3.7104
DW	1.1166

Analysis of variance:

Source	DF	Sum of squares	Mean squares	F	Pr > F
Model	1	10626.5928	10626.5928	771.8802	< 0.0001
Error	5	68.8358	13.7672		
Corrected Total	6	10695.4286			

Computed against model Y=Mean(Y)

Equation of the model:
Temperature = 72.49818-3.68431*Altitude

Explained variance = R^2 = 0.9936; 0.9936 × 10695.4286 = 10626.59.

b.   Unexplained variance = 10695.4286 - 10626.59 = 68.84.

c. $n = 7$; Standard error of the estimate = 3.7104; $\bar{x}$= 20.1429; Σx = 141; Σx^2 = 3623. Using Excel's T.INV.2T function, the t critical value =**T.INV.2T(0.05,5)** = 2.5706.

Prediction interval: $38.0°F < y < 60.4°F$

Because the P-value < 0.05, there is sufficient evidence to support a claim of a linear correlation. It is reasonable to use the regression equation when making predictions.

19. a.   Enter the redshift and distance data in an Excel worksheet and be sure to include the labels. Select the **XLSTAT** add-in. Select **Modeling data**. Select **Linear regression**. Enter the distance data range in the

Y/Dependent variables box. Enter the redshift data in the **X/Explanatory variables** box. Click **Sheet**. Click **Variables labels**. Click the **Outputs** tab. Click **Descriptive statistics** and **Analysis of variance**. Click **OK**.

Goodness of fit statistics:					
Observations	6.0000				
Sum of weights	6.0000				
DF	4.0000				
R²	1.0000				
Adjusted R²	1.0000				
MSE	0.0000				
RMSE	0.0014				
DW	1.1957				

Analysis of variance:					
Source	DF	Sum of squares	Mean squares	F	Pr > F
Model	1	0.4663	0.4663	253411.6887	< 0.0001
Error	4	0.0000	0.0000		
Corrected Total	5	0.4663			
Computed against model Y=Mean(Y)					

Equation of the model:					
Distance = -0.00440+13.99990*Redshift					

Explained variance = R^2 = 0.999984; 0.999984× 0.466283 = 0.466276.

b. Unexplained variance = 0.466283 − 0.433276= 0.000007

c. $n = 6$; Standard error of the estimate = 0.0014; $\bar{x}$ = 0.0404; Σx = 0.2426; Σx^2 = 0.012188. Using Excel's T.INV.2T function, the t critical value =**T.INV.2T(0.10,4)** = 2.1318.

Prediction interval: 0.168 billion light-years $< y <$ 0.176 billion light-years

Because the P-value < 0.10, there is sufficient evidence to support a claim of a linear correlation. It is reasonable to use the regression equation when making predictions.

Section 10-4, Beyond the Basics

21. $58.9 < \beta_0 < 103$; $2.46 < \beta_1 < 3.98$

$$\text{CI for } \beta_0$$

$$b_0 - E < \beta_0 < b_0 + E$$

$$b_0 = 80.93; \ E = t_{\alpha/2} s_e \sqrt{\frac{1}{n} + \frac{\bar{x}^2}{\Sigma x^2 - \frac{(\Sigma x)^2}{n}}} = 2.024 (5.94376)\sqrt{\frac{1}{40} + \frac{29.02^2}{33933 - \frac{1160.7^2}{40}}} = 22.06$$

$$\text{CI for } \beta_1$$

$$b_1 - E < \beta_1 < b_1 + E$$

$$b_1 = 3.2186; \ E = t_{\alpha/2} \cdot \frac{s_e}{\sqrt{\Sigma x^2 - \frac{(\Sigma x)^2}{n}}} = 2.024 \cdot \frac{5.94376}{\sqrt{33933 - \frac{1160.7^2}{40}}} = 0.757$$

Section 10-5, Basic Skills and Concepts

1. The response variable is weight and the predictor variables are length and chest size.

3. The unadjusted R^2 increases (or remains the same) as more variables are included, but the adjusted R^2 is adjusted for the number of variables and sample size. The unadjusted R^2 incorrectly suggests that the best multiple regression equation is obtained by including all of the available variables, but by taking into account the sample size and number of predictor variables, the adjusted R^2 is much more helpful in weeding out variables that should not be included.

5. LDL $= 47.4 + 0.085$ WT $+ 0.497$ SYS.

7. No. The P-value of 0.149 is not very low, and the values of R^2 (0.098) and adjusted R^2 (0.049) are not high. Although the multiple regression equation fits the sample data best, it is not a good fit.

9. HWY (highway fuel consumption) because it has the best combination of small P-value (0.000) and highest adjusted R^2 (0.920).

11. CITY $= -3.15 + 0.819$ HWY. That equation has a low P-value of 0.000 and its adjusted R^2 value of 0.920 isn't very much less than the values of 0.928 and 0.935 that use two predictor variables, so in this case it is better to use the one predictor variable instead of two.

13. Open the **CIGARET** data file on your data disk. For the first analysis, the x variable is FLTar and the y variable is FLNic. Select the **XLSTAT** add-in. Select **Modeling data**. Select **Linear regression**. Enter the FLNic data range in the **Y/Dependent variables** box. Enter the FLTar data range in the **X/Explanatory variables** box. Click **Sheet**. Click **Variables labels**. Click the **Outputs** tab. Click **Correlations** and **Analysis of variance**. Click **OK**.

Goodness of fit statistics:

Observations	25.0000
Sum of weights	25.0000
DF	23.0000
R²	0.8819
Adjusted R²	0.8768
MSE	0.0076
RMSE	0.0870
DW	1.4093

Analysis of variance:

Source	DF	Sum of squares	Mean squares	F	Pr > F
Model	1	1.2996	1.2996	171.7862	< 0.0001
Error	23	0.1740	0.0076		
Corrected Total	24	1.4736			

Computed against model Y=Mean(Y)

Equation of the model:
FLNic = 0.08000+0.06333*FLTar

For the second analysis, the x variable is FLCO and the y variable is FLNic. Select the **XLSTAT** add-in. Select **Modeling data**. Select **Linear regression**. Enter the FLNic data range in the **Y/Dependent variables** box. Enter the FLCO data range in the **X/Explanatory variables** box. Click **Sheet**. Click **Variables labels**. Click the **Outputs** tab. Click **Correlations** and **Analysis of variance**. Click **OK**.

Goodness of fit statistics:					
Observations	25.0000				
Sum of weights	25.0000				
DF	23.0000				
R²	0.4604				
Adjusted R²	0.4369				
MSE	0.0346				
RMSE	0.1859				
DW	1.5887				

Analysis of variance:

Source	DF	Sum of squares	Mean squares	F	Pr > F
Model	1	0.6784	0.6784	19.6235	0.0002
Error	23	0.7952	0.0346		
Corrected Total	24	1.4736			

Computed against model Y=Mean(Y)

Equation of the model:
FLNic = 0.32813+0.03972*FLCO

For the third analysis, the x variables are FLTar and FLCO, and the y variable is FLNic. The x variables need to be in adjacent columns of the Excel worksheet. Select the **XLSTAT** add-in. Select **Modeling data**. Select **Linear regression**. Enter the FLNic data range in the **Y/Dependent variables** box. Enter the data range of the two x variables in the **X/Explanatory variables** box. Click **Sheet**. Click **Variables labels**. Click the **Outputs** tab. Click **Correlations** and **Analysis of variance**. Click **OK**.

Goodness of fit statistics:					
Observations	25.0000				
Sum of weights	25.0000				
DF	22.0000				
R²	0.9328				
Adjusted R²	0.9267				
MSE	0.0045				
RMSE	0.0671				
DW	1.7787				

Analysis of variance:

Source	DF	Sum of squares	Mean squares	F	Pr > F
Model	2	1.3745	0.6873	152.6141	< 0.0001
Error	22	0.0991	0.0045		
Corrected Total	24	1.4736			

Computed against model Y=Mean(Y)

Equation of the model:
FLNic = 0.12714+0.08780*FLTar-0.02500*FLCO

The best regression equation is $\hat{y} = 0.127 + 0.0878x_1 - 0.0250x_2$, where x_1 represents tar and x_2 represents carbon monoxide. It is best because it has the highest adjusted R^2 value of 0.927 and a P-value <0.0001. It is a good regression equation for predicting nicotine content because it has a high value of adjusted R^2 and a low P-value.

15. Open the **IQBRAIN** data file on your data disk. For the first analysis, the x variable is VOL and the y variable is IQ. Select the **XLSTAT** add-in. Select **Modeling data**. Select **Linear regression**. Enter the IQ data range in the **Y/Dependent variables** box. Enter the VOL data range in the **X/Explanatory variables** box. Click **Sheet**. Click **Variables labels**. Click the **Outputs** tab. Click **Correlations** and **Analysis of variance**. Click **OK**.

Goodness of fit statistics:					
Observations	20.0000				
Sum of weights	20.0000				
DF	18.0000				
R²	0.0040				
Adjusted R²	-0.0513				
MSE	183.4819				
RMSE	13.5455				
DW	0.7830				

Analysis of variance:

Source	DF	Sum of squares	Mean squares	F	Pr > F
Model	1	13.3255	13.3255	0.0726	0.7906
Error	18	3302.6745	183.4819		
Corrected Total	19	3316.0000			

Computed against model Y=Mean(Y)

Equation of the model:
IQ = 108.54721-0.00670*VOL

For the second analysis, the *x* variable is WT and the *y* variable is IQ. Select the **XLSTAT** add-in. Select **Modeling data**. Select **Linear regression**. Enter the IQ data range in the **Y/Dependent variables** box. Enter the WT data range in the **X/Explanatory variables** box. Click **Sheet**. Click **Variables labels**. Click the **Outputs** tab. Click **Correlations** and **Analysis of variance**. Click **OK**. The output is displayed below.

For the third analysis, the *x* variables are VOL and WT, and the *y* variable is IQ. The *x* variables need to be in adjacent columns of the Excel worksheet. Select the **XLSTAT** add-in. Select **Modeling data**. Select **Linear regression**. Enter the IQ data range in the **Y/Dependent variables** box. Enter the data range of the two *x* variables in the **X/Explanatory variables** box. Click **Sheet**. Click **Variables labels**. Click the **Outputs** tab. Click **Correlations** and **Analysis of variance**. Click **OK**. The output is displayed below.

The best regression equation is $\hat{y} = 109 - 0.00670x_1$, where x_1 represents volume. It is best because it has the highest adjusted R^2 value of -0.0513 and the lowest *P*-value of 0.791. The three regression equations all have adjusted values of R^2 that are very close to 0, so none of them are good for predicting IQ. It does not appear that people with larger brains have higher IQ scores.

Second analysis output:

Goodness of fit statistics:					
Observations	20.0000				
Sum of weights	20.0000				
DF	18.0000				
R²	0.0000				
Adjusted R²	-0.0555				
MSE	184.2209				
RMSE	13.5728				
DW	0.7893				

Analysis of variance:

Source	DF	Sum of squares	Mean squares	F	Pr > F
Model	1	0.0242	0.0242	0.0001	0.9910
Error	18	3315.9758	184.2209		
Corrected Total	19	3316.0000			

Computed against model Y=Mean(Y)

Equation of the model:
IQ = 101.13860-0.00178*WT

Third analysis output:

Goodness of fit statistics:					
Observations	20.0000				
Sum of weights	20.0000				
DF	17.0000				
R²	0.0041				
Adjusted R²	-0.1130				
MSE	194.2526				
RMSE	13.9375				
DW	0.7866				

Analysis of variance:					
Source	DF	Sum of squares	Mean squares	F	Pr > F
Model	2	13.7060	6.8530	0.0353	0.9654
Error	17	3302.2940	194.2526		
Corrected Total	19	3316.0000			
Computed against model Y=Mean(Y)					

Equation of the model:
IQ = 108.25664-0.00694*VOL+0.00722*WT

Section 10-5, Beyond the Basics

17. For H_0: $\beta_1 = 0$, the test statistic is $t = \frac{0.7072}{0.1289} = 5.486$, the P-value is 0.000, and the critical values are $t = \pm 2.110$, so reject H_0 and conclude that the regression coefficient of $b_1 = 0.707$ should be kept. For H_0: $\beta_2 = 0$, the test statistic is $t = \frac{0.1636}{0.1266} = 1.292$, the P-value is 0.213, and the critical values are $t = \pm 2.110$, so fail to reject H_0 and conclude that the regression coefficient of $b_2 = 0.164$ should be omitted. It appears that the regression equation should include the height of the mother as a predictor variable, but the height of the father should be omitted.

19. Open the **BEARS** data file on your data disk. Recode the SEX variable into a dummy variable of sex following the instructions in Exercise 19. For the analysis, the x variables are AGE and dummy variable of sex, and the y variable is WEIGHT. The x variables need to be in adjacent columns of the Excel worksheet. Select the **XLSTAT** add-in. Select **Modeling data**. Select **Linear regression**. Enter the WEIGHT data range in the **Y/Dependent variables** box. Enter the data range of the two x variables in the **X/Explanatory variables** box. Click **Sheet**. Click **Variables labels**.

Equation of the model:			
WEIGHT = 3.06237+82.37938*DUMMY-SEX+2.90526*AGE			

$\hat{y} = 3.06 + 82.4x_1 + 2.91x_2$, where x_1 represents age and x_2 represents age.

a.   Female: $\hat{y} = 3.06 + 82.4(0) + 2.91(20) = 61$ lb .

b.   Male: $\hat{y} = 3.06 + 82.4(1) + 2.91(20) = 144$ lb .

The sex of the bear does appear to have an effect on its weight. The regression equation indicates that the predicted weight of a male bear is about 82 lb more than the predicted weight of a female bear with other characteristics being the same.

Section 10-6, Basic Skills and Concepts

1.   Since the area of a square is the square of its side, the best model is $y = x^2$; quadratic; $R^2 = 1$.

3.   10.3% of the variation in Super Bowl points can be explained by the quadratic model that relates the variable of year and the variable of points scored. Because such a small percentage of the variation is explained by the model, the model is not very useful.

5.   Enter the *t* and *d* data in an Excel worksheet. Click and drag over the *t* and *d* data range to select this range for the chart. Click **INSERT** at the top of the screen. Select **Scatter** in the Charts group. Select the leftmost figure in the top row. Right-click on a plotted point. Select **Add Trendline**. Click **Exponential, Display Equation on Chart**, and **Display R-squared value on chart**. To display charts of the other models, right-click on a plotted point and select **Add** or **Format Trendline**. Then select, one at a time, **Linear, Logarithmic, Polynomial Order 2** (same as quadratic), and **Power**. For each model, click **Display Equation on Chart** and **Display R-squared value** on chart. If the trend line from the preceding model is displayed on the graph, right-click on the trend line and select Delete.

Exponential model:

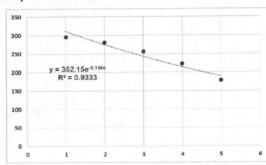

Linear model:

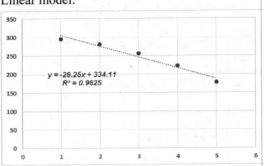

Logarithmic model:

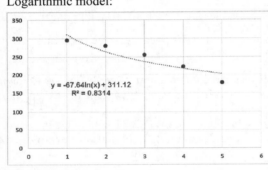

Polynomial order 2 (or Quadratic) model:

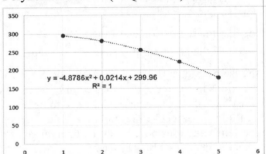

Power model:

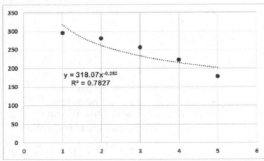

With $R^2 = 1$, the quadratic model best fits the data.

7.   Enter the year and value data in an Excel worksheet. Click and drag over the year and value data range to select this range for the chart. Click **INSERT** at the top of the screen. Select **Scatter** in the Charts group. Select the leftmost figure in the top row. Right-click on a plotted point. Select **Add Trendline**. Click **Linear, Display Equation on Chart**, and **Display R-squared value on chart**. To display charts of the other models, first right-click on the trend line and select **Delete**. Then right-click on a plotted point and select **Add Trendline**. Then

select, one at a time, **Linear**, **Logarithmic**, **Polynomial Order 2** (same as quadratic), and **Power**. For each model, click **Display Equation on Chart** and **Display R-squared value** on chart.

Exponential model:

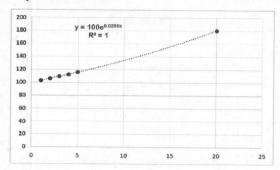

Linear model:

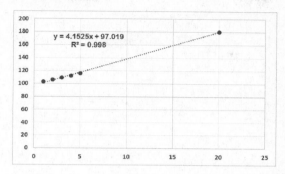

Logarithmic model:

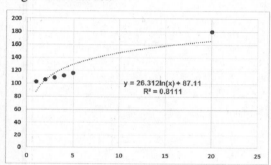

Polynomial Order 2 (same as Quadratic) model:

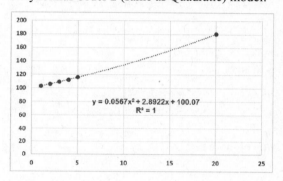

Power model:

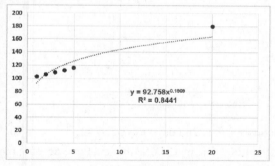

With $R^2 = 1$, the exponential model best fits the data.

9. Enter the days (1 through 21) and amounts grossed in an Excel worksheet. Click and drag over the days and amount grossed data range to select this range for the chart. Click **INSERT** at the top of the screen. Select **Scatter** in the Charts group. Select the leftmost figure in the top row. Right-click on a plotted point. Select **Add Trendline**. Click **Linear**, **Display Equation on Chart**, and **Display R-squared value on chart**. To display charts of the other models, first right-click on the trend line and select **Delete**. Then right-click on a plotted point and select **Add Trendline**. Then select, one at a time, **Linear**, **Logarithmic**, **Polynomial Order 2** (same as quadratic), and **Power**. For each model, click **Display Equation on Chart** and **Display R-squared value** on chart.

Exponential model:

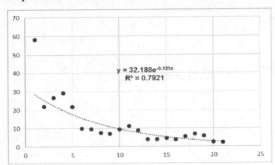

Linear model:

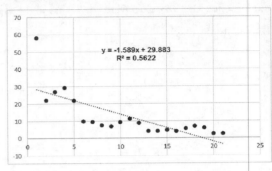

Logarithmic model:

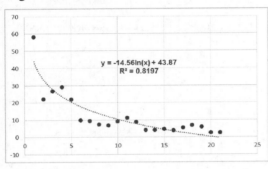

Polynomial Order 2 (same as Quadratic) model:

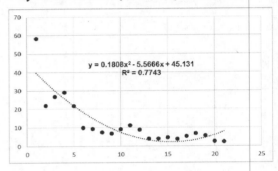

Power model:

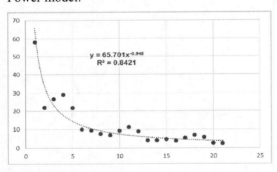

With $R^2 = 0.842$, the power model best fits the data. Prediction for the 22nd day: $y = 65.7(22)^{-0.945} = \$3.5$ million, which isn't very close to the actual amount of $2.2 million. The model does not take into account the fact that movies do better on weekend days.

11. Enter the TNT and Richter scale data in an Excel worksheet. Click and drag over the data range to select this range for the chart. Click **INSERT** at the top of the screen. Select **Scatter** in the Charts group. Select the leftmost figure in the top row. Right-click on a plotted point. Select **Add Trendline**. Click **Linear**, **Display Equation on Chart**, and **Display R-squared value on chart**. To display charts of the other models, first right-click on the trend line and select **Delete**. Then right-click on a plotted point and select **Add Trendline**. Then select, one at a time, **Linear**, **Logarithmic**, **Polynomial Order 2** (same as quadratic), and **Power**. For each model, click **Display Equation on Chart** and **Display R-squared value** on chart.

Exponential model:

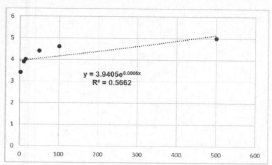

Linear model:

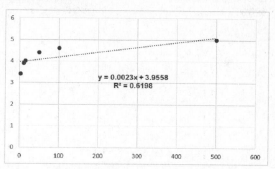

Logarithmic model:

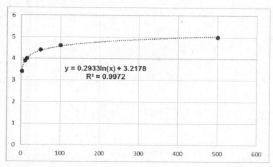

Polynomial Order 2 (same as Quadratic) model:

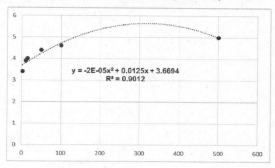

Power model:

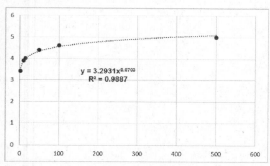

With $R^2 = 0.997$, the logarithmic model best fits the data.

13. Enter the day (1 through 7) and number of bacteria data in an Excel worksheet. Click and drag over the data range to select this range for the chart. Click **INSERT** at the top of the screen. Select **Scatter** in the Charts group. Select the leftmost figure in the top row. Right-click on a plotted point. Select **Add Trendline**. Click **Linear**, **Display Equation on Chart**, and **Display R-squared value on chart**. To display charts of the other models, first right-click on the trend line and select **Delete**. Then right-click on a plotted point and select **Add Trendline**. Then select, one at a time, **Linear**, **Logarithmic**, **Polynomial Order 2** (same as quadratic), and **Power**. For each model, click **Display Equation on Chart** and **Display R-squared value** on chart.

Exponential model:

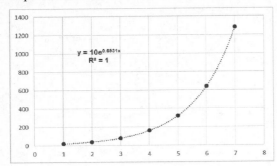

Linear model:

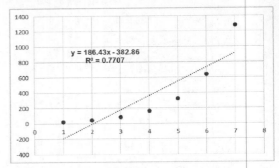

Logarithmic model:

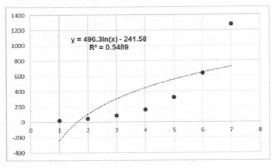

Polynomial Order 2 (same as Quadratic) model:

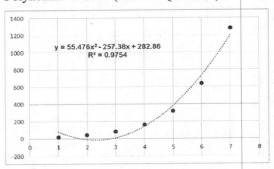

Power model:

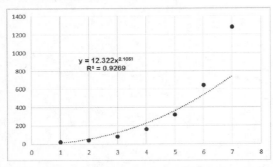

With $R^2 = 1$, the exponential model best fits the data.

15. Enter the year (1 through 10) and the Dow Jones Industrial Average data in an Excel worksheet. Click and drag over the data range to select this range for the chart. Click **INSERT** at the top of the screen. Select **Scatter** in the Charts group. Select the leftmost figure in the top row. Right-click on a plotted point. Select **Add Trendline**. Click **Linear**, **Display Equation on Chart**, and **Display R-squared value on chart**. To display charts of the other models,first right-click on the trend line and select **Delete**. Then right-click on a plotted point and select **Add Trendline**. Then select, one at a time, **Linear**, **Logarithmic**, **Polynomial Order 2** (same as quadratic), and **Power**. For each model, click **Display Equation on Chart** and **Display R-squared value** on chart.

Exponential model:

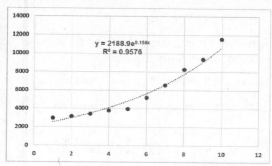

Linear model:

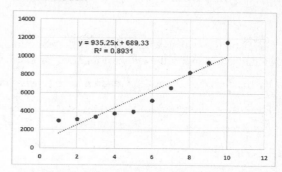

Logarithmic model:

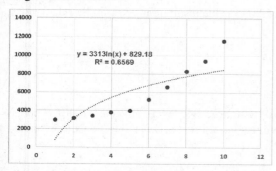

Polynomial Order 2 (same as Quadratic) model:

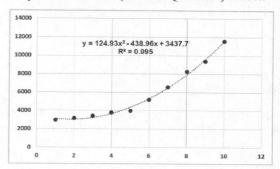

Power model:

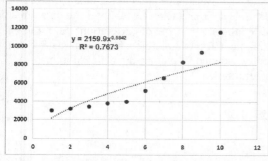

With $R^2 = 0.995$, the quadratic model best fits the data The projected value for 2010 is

$y = 125(21)^2 - 439(21) + 3438 = 49{,}344$ (Tech: 49,312), which is dramatically greater than the actual value of 11,655.

Section 10-6, Beyond the Basics

17. a. Exponential: $y = 2^{\frac{2}{3}(x-1)}$ [or $y = 0.629961(1.587401)^x$ for an initial value of 1 that doubles every 1.5 years].

b. Exponential: $y = 1.36558(1.42774)^x$, where 1971 is coded as 1.

c. Moore's law does appear to be working reasonably well. With $R^2 = 0.990$, the model appears to be very good.

Chapter Quick Quiz

1. $r = \pm 0.878$

2. Based on the critical values of ± 0.878 (assuming a 0.05 significance level), conclude that there is not sufficient evidence to support the claim of a linear correlation between systolic and diastolic readings.

3. The best predicted diastolic reading is 90.6, which is the mean of the five sample diastolic readings.

4. The best predicted diastolic reading is $\hat{y} = -1.99 + 0.698(125) = 85.3$, which is found by substituting 125 for x in the regression equation.

5. $r^2 = 0.342$

6. False; there could be another relationship.

7. False, correlation does not imply causation.

8. $r = 1$

9. Because r must be between -1 and 1 inclusive, the value of 3.335 is the result of an error in the calculations.

10. $r = -1$

Review Exercises

1. a. $r = 0.926$. Critical values: $r = \pm 0.707$ (assuming a 0.05 significance level). P-value = 0.001. There is sufficient evidence to support the claim that there is a linear correlation between duration and interval-after time.

    b. $r^2 = (0.926)^2 = 0.857 = 85.7\%$ c. $\hat{y} = 34.8 + 0.234x$

2. a. Enter the height and interval-after times in an Excel worksheet. Be sure to include the labels. Select the **XLSTAT** add-in. Select **Correlation/Association tests**. Select **Correlation tests**. Enter the data range in the Observations/variables table box. Check the **Variable labels** box. Type of correlation: **Pearson**. Click the **Outputs** tab. Select **Correlations** and **p-values**. Click the **Charts** tab. Select **Scatter plots**. Click **OK**. The scatterplot suggests that there is not sufficient sample evidence to support the claim of a linear correlation between heights of eruptions and interval-after times.

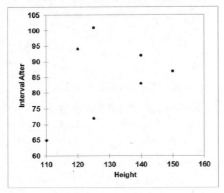

    b. r = 0.2694. Critical values: $r = \pm 0.707$ (assuming a 0.05 significance level). P-value = 0.5188. There is not sufficient evidence to support the claim that there is a linear correlation between height and interval-after time.

Correlation matrix (Pearson):		
Variables	Height	Interval After
Height	1	0.2694
Interval After	0.2694	1

p-values:		
Variables	Height	Interval After
Height	0	0.5188
Interval After	0.5188	0

c. Select the **XLSTAT** add-in. Select **Modeling data**. Select **Linear regression**. Enter the interval after data range in the **Y/Dependent variables** box. Enter the height data range in the **X/Explanatory variables** box. Click **Sheet**. Click **Variable labels**. Click **OK**.

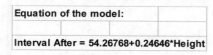

Equation of the model:	
Interval After = 54.26768+0.24646*Height	

d. $\hat{y} = 54.3 + 0.246(100) = 78.9$ min

3. a. Enter the duration and height data in an Excel worksheet. Be sure to include the labels. Select the **XLSTAT** add-in. Select **Correlation/Association tests**. Select **Correlation tests**. Enter the data range in the Observations/variables table box. Check the **Variable labels** box. Type of correlation: **Pearson**. Click the **Outputs** tab. Select **Correlations** and **p-values**. Click the **Charts** tab. Select **Scatter plots**. Click **OK**. The scatterplot suggests that there is not sufficient sample evidence to support the claim of a linear correlation between duration and height.

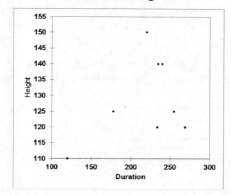

b. $r = 0.389$. Critical values: $r = \pm 0.707$ (assuming a 0.05 significance level). P-value = 0.3403. There is not sufficient evidence to support the claim that there is a linear correlation between duration and height.

Correlation matrix (Pearson):		
Variables	Duration	Height
Duration	1	0.3894
Height	0.3894	1

p-values:		
Variables	Duration	Height
Duration	0	0.3403
Height	0.3403	0

c. Select the **XLSTAT** add-in. Select **Modeling data**. Select **Linear regression**. Enter the height data range in the **Y/Dependent variables** box. Enter the duration range in the **X/Explanatory variables** box. Click **Sheet**. Click **Variable labels**. Click **OK**.

Equation of the model:	
Height = 105.19181+0.10763*Duration	

d. The regression line does not fit the points well, so the best predicted height is the mean; $\overline{y} = 128.8$ ft.

4. Enter the time and height data in an Excel worksheet. Be sure to include labels. Select the **XLSTAT** add-in. Select **Correlation/Association tests**. Select **Correlation tests**. Enter the data range in the Observations/variables table box. Check the **Variable labels** box. Type of correlation: **Pearson**. Click the **Outputs** tab. Select **Correlations** and **p-values**. Click the **Charts** tab. Select **Scatter plots**. Click **OK**. $r = 0.450$. Critical values: $r = \pm 0.632$ (assuming a 0.05 significance level). P-value = 0.1918. There is not sufficient evidence to support the claim that there is a linear correlation between time and height. Although there is no linear correlation between time and height, the scatterplot shows a very distinct pattern revealing that time and height are associated by some function that is not linear.

Correlation matrix (Pearson):

Variables	Time	Height
Time	1	0.4501
Height	0.4501	1

p-values:

Variables	Time	Height
Time	0	0.1918
Height	0.1918	0

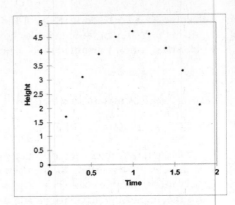

5.   Enter the data in an Excel worksheet. Be sure to include the labels. Select the **XLSTAT** add-in. Select **Modeling data**. Select **Linear regression**. Enter the interval after data range in the **Y/Dependent variables** box. Enter the duration and interval before data range in the **X/Explanatory variables** box. Click **Sheet**. Click **Variable labels**. Click the **Outputs** tab. Select **Correlations** and **Analysis of variance**. Click **OK**. Interval After $= 50.1 + 0.242$ Duration $- 0.178$ Interval Before, or $\hat{y} = 50.1 + 0.242x_1 - 0.178x_2$. $R^2 = 0.8716$; adjusted $R^2 = 0.8202$; P-value $= 0.0059$. With high values of R^2 and adjusted R^2 and a small P-value of 0.006, it appears that the regression equation can be used to predict the time interval after an eruption given the duration of the eruption and the time interval before that eruption.

Goodness of fit statistics:

Observations	8.0000
Sum of weights	8.0000
DF	5.0000
R²	0.8716
Adjusted R²	0.8202
MSE	26.6034
RMSE	5.1578
DW	2.5924

Analysis of variance:

Source	DF	Sum of squares	Mean squares	F	Pr > F
Model	2	902.9830	451.4915	16.9712	0.0059
Error	5	133.0170	26.6034		
Corrected Total	7	1036.0000			

Computed against model Y=Mean(Y)

Equation of the model:

Interval After = 50.09002+0.24179*Duration-0.17790*Interval Before

Cumulative Review Exercises

1.   Enter the before and after data in an Excel worksheet. Use a formula to calculate a before-after difference for each subject. Select the **XLSTAT** add-in. Select **Describing data**. Select **Descriptive statistics**. Enter the range of the differences in the Quantitative data window. Click the **Outputs** tab. Select **Mean** and **Standard deviation (n-1)**. Click **OK**. $\bar{x} = 3.3$ lb, $s = 5.7$ lb.

Descriptive statistics (Quantitative data)

Statistic	Difference
Mean	3.2500
Standard deviation (n-1)	5.7009

2.   Enter the before data in an Excel worksheet. Select the **XLSTAT** add-in. Select **Preparing data**. Select **Variables transformation**. Enter the range of the before data in the data box. Select the **Standardize (n-1)** transformation. Click **OK**. The highest weight before the diet is 212 lb, which converts to
$$z = \frac{212 - 179.4}{21.0} = 1.55.$$ The highest weight is not unusual because its z score of 1.55 shows that it is within 2 standard deviations of the mean.

Transformed data:	
Before	
0.1723	
1.5508	
-0.1129	
1.4082	
-1.1587	
-0.8259	
-0.5883	
-0.4456	

3. Enter the before and after data in an Excel worksheet. Select the **XLSTAT** add-in. Select **Parametric tests**. Select **Two-sample *t*-test and *z*-test**. Enter the before data range in the Sample 1 box. Enter the after data range in the Sample 2 box. Select **Paired samples**. Select **Student's t test**. Click the **Options** tab. Alternative hypothesis: **Mean 1 – Mean 2 > D**. Hypothesized difference (D): **0**. Significance level (%): **5**. Click **OK**. $H_0 : \mu_d = 0$. $H_0 : \mu_d > 0$. Test statistic: $t = 1.613$. Critical value: $t = 1.895$. P-value = 0.0754. Fail to reject H_0. There is not sufficient evidence to support the claim that the diet is effective.

t-test for two paired samples / Upper-tailed test:	
95% confidence interval on the difference between the means:	
(-0.5686 , +Inf)	
Difference	3.2500
t (Observed value)	1.6125
t (Critical value)	1.8946
DF	7
p-value (one-tailed)	0.0754
alpha	0.05

4. Enter the before data in an Excel worksheet. Select the **XLSTAT** add-in. Select **Describing data**. Select **Descriptive statistics**. Enter the range of the before data in the Quantitative data box. Click the **Options** tab. Set the confidence interval (%) at **95**. Click the **Outputs** tab. Select **Mean, Lower bound on mean (95%)**, and **Upper bound on mean (95%)**. Click **OK**. The 95% confidence interval is $161.8 \text{ lb} < \mu < 197.0 \text{ lb}$. We have 95% confidence that the interval limits of 161.8 lb and 197.0 lb contain the true value of the mean of the population of all subjects before the diet.

Descriptive statistics (Quantitative data):	
Statistic	**Before**
Mean	179.3750
Lower bound on mean (95%)	161.7756
Upper bound on mean (95%)	196.9744

5. a. Enter the before and after data in an Excel worksheet. Be sure to include labels. Select the **XLSTAT** add-in. Select **Correlation/Association tests**. Select **Correlation tests**. Enter the data range in the Observations/variables table box. Check the **Variable labels** box. Type of correlation: **Pearson**. Click the **Outputs** tab. Select **Correlations** and **p-values**. $r = 0.9648$. Critical values: $r = \pm 0.707$ (assuming a 0.05 significance level). P-value = 0.0001. There is sufficient evidence to support the claim that there is a linear correlation between before and after weights.

Correlation matrix (Pearson):				p-values:		
Variables	**Before**	**After**		**Variables**	**Before**	**After**
Before	1	0.9648		Before	0	0.0001
After	0.9648	1		After	0.0001	0

b. $r = 1$

    c. $r = 1$

    d.   The effectiveness of the diet is determined by the amounts of weight lost, but the linear correlation coefficient is not sensitive to different amounts of weight loss. Correlation is not a suitable tool for testing the effectiveness of the diet.

6. a.   Using Excel's NORM.DIST function, the percentage of babies born with a weight greater than 3500 g =**1-NORM.DIST(3500,3420,495,TRUE)** = 0.4358 or 43.58%.

    b.   Using Excel's NORM.INV function, the 10th percentile =**NORM.INV(0.10,3420,495)** = 2785.6 g.

    c.   Using Excel's NORM.DIST function, the percentage of babies that are less than 2450 g =**NORM.DIST(2450,3420,495,TRUE)** = 0.0250 OR 2.50%. The percentage of babies that are more than 4390 g =1-NORM.DIST(4390,3420,495,TRUE) = 0.0250 OR 2.50%. 2.50% + 2.50% = 5.00%. Yes, many of the babies do require special treatment.

7. a. $H_0: p = 0.5$. $H_1: p > 0.5$. Select the **XLSTAT** add-in. Select **Parametric** tests. Select **Tests for one proportion**. Select **Frequency** for the data format. Frequency: **269**, Sample size: **456**, Test proportion: **0.5**. Select **z test**. Click the **Options** tab. Alternative hypothesis: **Proportion – Test proportion > D**, Hypothesized difference (D): **0**, Significance level (%): **5**. Click **OK**. Test statistic: $z = 3.84$. Critical value: $z = 1.645$. P-value < 0.0001. Reject H_0. There is sufficient evidence to support the claim that the majority of us say that honesty is always the best policy.

z-test for one proportion / Upper-tailed test:		
Difference	0.0899	
z (Observed value)	3.8400	
z (Critical value)	1.6449	
p-value (one-tailed)	< 0.0001	
alpha	0.05	

    b.   The sample is a voluntary response (or self-selected) sample. This type of sample suggests that the results given in part (a) are not necessarily valid.

8. a.   Nominal

    b.   Ratio

    c.   Discrete

    d. $\dfrac{304}{529} = 0.575$

    e.   Parameter

9. a. $\left(\dfrac{304}{529}\right)^2 = 0.330$

    b. $\dfrac{304 + 156}{529} = \dfrac{460}{529} = 0.870$

    c. $\dfrac{514}{529} = 0.972$

    d. $\dfrac{39}{529} = 0.0737 = 7.37\%$

10.

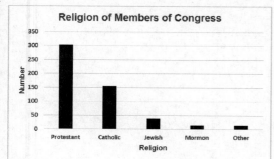

Chapter 11

Goodness-of-Fit and Contingency Tables

Section 11-2, Basic Skills and Concepts

1. The test is to determine whether the observed frequency counts agree with the claimed uniform distribution so that frequencies for the different days are equally likely.

3. Because the given frequencies differ substantially from frequencies that are all about the same, the χ^2 test statistic should be large and the P-value should be small.

5. Enter the observed frequency data in an Excel worksheet. The expected proportion for each day of the week is 1/7 or 0.1429. Enter this expected proportion (or the formula =1/7) for each day of the week. Select the **XLSTAT** add-in. Select **Parametric tests**. Select **Multinomial good of fit test**. In the **Frequencies** box, enter the range of cells containing the observed frequencies. For **Data format**, click **Proportions**. In the **Expected** proportions box, enter the range of cells containing the expected proportions. If you included column labels in the data ranges, click **Column labels**. Click **Chi-square test**. For the **Significance level (%)**, enter **5**. Click **OK**.

Chi-square test:	
Chi-square (Observed value)	1934.9791
Chi-square (Critical value)	12.5916
DF	6
p-value	< 0.0001
alpha	0.05

Test statistic: $\chi^2 = 1934.9791$. Critical value: $\chi^2 = 12.5916$. P-value < 0.0001. There is sufficient evidence to warrant rejection of the claim that the days of the week are selected with a uniform distribution with all days having the same chance of being selected.

7. Using Excel's CHISQ.INV.RT function, the critical value =**CHISQ.INV.RT(0.05,9)** = 16.9190. Using Excel's CHISQ.DIST.RT function, the P-value =**CHISQ.DIST.RT(8.815,9)** = 0.5156. There is not sufficient evidence to warrant rejection of the claim that the observed outcomes agree with the expected frequencies. The slot machine appears to be functioning as expected.

9. Enter the observed frequency data in an Excel worksheet. The expected proportion for each month is 1/12 or 0.0833. Enter this expected proportion (or the formula =1/12) for each month. Select the **XLSTAT** add-in. Select **Parametric tests**. Select **Multinomial good of fit test**. In the **Frequencies** box, enter the range of cells containing the observed frequencies. For **Data format**, click **Proportions**. In the **Expected** proportions box, enter the range of cells containing the expected proportions. If you included column labels in the data ranges, click **Column labels**. Click **Chi-square test**. For the **Significance level (%)**, enter **5**. Click **OK**.

Chi-square test:	
Chi-square (Observed value)	10.3750
Chi-square (Critical value)	19.6751
DF	11
p-value	0.4970
alpha	0.05

Test statistic: $\chi^2 = 10.3750$. Critical value: $\chi^2 = 19.6751$. P-value = 0.4970. There is not sufficient evidence to warrant rejection of the claim that homicides in New York City are equally likely for each of the 12 months. There is not sufficient evidence to support the police commissioner's claim that homicides occur more often in the summer when the weather is better.

11. Enter the observed frequency data in an Excel worksheet. The expected proportion for each die is 1/6 or 0.1667. Enter this expected proportion (or the formula =1/6) for each die. Select the **XLSTAT** add-in. Select **Parametric tests**. Select **Multinomial good of fit test**. In the **Frequencies** box, enter the range of cells containing the observed frequencies. For **Data format**, click **Proportions**. In the **Expected** proportions box, enter the range of cells containing the expected proportions. If you included column labels in the data ranges, click **Column labels**. Click **Chi-square test**. For the **Significance level (%)**, enter **5**. Click **OK**.

Chi-square test:	
Chi-square (Observed value)	5.8600
Chi-square (Critical value)	11.0705
DF	5
p-value	0.3201
alpha	0.05

Test statistic: $\chi^2 = 5.8600$. Critical value: $\chi^2 = 11.0705$. P-value = 0.3201. There is not sufficient evidence to support the claim that the outcomes are not equally likely. The outcomes appear to be equally likely, so the loaded die does not appear to behave differently from a fair die.

13. Enter the observed frequency data in an Excel worksheet. The expected proportion for each post position is 1/10 or 0.1. Enter this expected proportion for each post position. Select the **XLSTAT** add-in. Select **Parametric tests**. Select **Multinomial good of fit test**. In the **Frequencies** box, enter the range of cells containing the observed frequencies. For **Data format**, click **Proportions**. In the **Expected** proportions box, enter the range of cells containing the expected proportions. If you included column labels in the data ranges, click **Column labels**. Click **Chi-square test**. For the **Significance level (%)**, enter **5**. Click **OK**.

Chi-square test:	
Chi-square (Observed value)	13.4828
Chi-square (Critical value)	16.9190
DF	9
p-value	0.1420
alpha	0.05

Test statistic: $\chi^2 = 13.4828$. Critical value: $\chi^2 = 16.9190$. *P*-value = 0.1420. There is not sufficient evidence to warrant rejection of the claim that the likelihood of winning is the same for the different post positions. Based on these results, post position should not be considered when betting on the Kentucky Derby race.

15. Enter the observed frequency data in an Excel worksheet. The expected proportion for each day is 1/7 or 0.1429. Enter this expected proportion (or the formula **=1/7**) for each day of the week. Select the **XLSTAT** add-in. Select **Parametric tests**. Select **Multinomial good of fit test**. In the **Frequencies** box, enter the range of cells containing the observed frequencies. For **Data format**, click **Proportions**. In the **Expected** proportions box, enter the range of cells containing the expected proportions. If you included column labels in the data ranges, click **Column labels**. Click **Chi-square test**. For the **Significance level (%)**, enter **1**. Click **OK**.

Chi-square test:	
Chi-square (Observed value)	29.8137
Chi-square (Critical value)	16.8119
DF	6
p-value	< 0.0001
alpha	0.01

Test statistic: $\chi^2 = 29.8137$. Critical value: $\chi^2 = 16.8119$. *P*-value < 0.0001. There is sufficient evidence to warrant rejection of the claim that the different days of the week have the same frequencies of police calls. The highest numbers of calls appear to fall on Friday and Saturday, and these are weekend days with disproportionately more partying and drinking.

17. Enter the observed frequency data in an Excel worksheet. Enter the expected proportions (or formulas e.g., **=2/16**) for each number of games played. Select the **XLSTAT** add-in. Select **Parametric tests**. Select **Multinomial good of fit test**. In the **Frequencies** box, enter the range of cells containing the observed frequencies. For **Data format**, click **Proportions**. In the **Expected** proportions box, enter the range of cells containing the expected proportions. If you included column labels in the data ranges, click **Column labels**. Click **Chi-square test**. For the **Significance level (%)**, enter **5**. Click **OK**.

Chi-square test:	
Chi-square (Observed value)	7.5786
Chi-square (Critical value)	7.8147
DF	3
p-value	0.0556
alpha	0.05

Test statistic: $\chi^2 = 7.5786$. Critical value: $\chi^2 = 7.8147$. *P*-value = 0.0556. There is not sufficient evidence to warrant rejection of the claim that the actual numbers of games fit the distribution indicated by the proportions listed in the given table.

19. Open the **M&M** data file on your data disk. You will first obtain the observed frequency of each color. Select **XLSTAT**. Select **Describing data**. Select **Descriptive statistics**. In the **Quantitative data** box, enter the range that encompasses all the data (**A1:F28**). Click the **Options tab**. Select **Descriptive statistics**. Click the **Outputs** tab. Select **No. of observations** and **No. of missing values**. Click **OK**. The observed frequencies equal the No. of observations minus No. of missing values.

Statistic	Red	Orange	Yellow	Brown	Blue	Green
No. of observations	27	27	27	27	27	27
No. of missing values	14	2	19	19	0	8
Observed frequencies	13	25	8	8	27	19

To use XLSTAT to solve the problem the data need to be entered in columns rather than rows. Enter the observed frequencies in a column of an Excel worksheet. Then enter the expected proportions in another column, being sure to match each proportion with its correct observed frequency. Select the **XLSTAT** add-in. Select **Parametric tests**. Select **Multinomial good of fit test**. In the **Frequencies** box, enter the range of cells containing the observed frequencies. For **Data format**, click **Proportions**. In the **Expected** proportions box, enter the range of cells containing the expected proportions. If you included column labels in the data ranges, click **Column labels**. Click **Chi-square test**. For the **Significance level (%)**, enter **5**. Click **OK**.

Chi-square test:	
Chi-square (Observed value)	6.6820
Chi-square (Critical value)	11.0705
DF	5
p-value	0.2454
alpha	0.05

Test statistic: $\chi^2 = 6.6820$. Critical value: $\chi^2 = 11.0705$. *P*-value = 0.2454. There is not sufficient evidence to warrant rejection of the claim that the color distribution is as claimed.

21. Enter the observed frequency data in an Excel worksheet. Enter the expected proportions (e.g., 0.301 for 30.1%). Select the **XLSTAT** add-in. Select **Parametric tests**. Select **Multinomial good of fit test**. In the **Frequencies** box, enter the range of cells containing the observed frequencies. For **Data format**, click **Proportions**. In the **Expected** proportions box, enter the range of cells containing the expected proportions. If you included column labels in the data ranges, click **Column labels**. Click **Chi-square test**. For the **Significance level (%)**, enter **1**. Click **OK**.

Chi-square test:	
Chi-square (Observed value)	3650.2514
Chi-square (Critical value)	20.0902
DF	8
p-value	< 0.0001
alpha	0.01

Test statistic: $\chi^2 = 3650.2514$. Critical value: $\chi^2 = 20.0902$. P-value < 0.0001. There is sufficient evidence to warrant rejection of the claim that the leading digits are from a population with a distribution that conforms to Benford's law. It does appear that the checks are the result of fraud (although the results cannot confirm that fraud is the cause of the discrepancy between the observed results and the expected results).

23. Enter the observed frequency data in an Excel worksheet. Enter the expected proportions (e.g., 0.301 for 30.1%). Select the **XLSTAT** add-in. Select **Parametric tests**. Select **Multinomial good of fit test**. In the **Frequencies** box, enter the range of cells containing the observed frequencies. For **Data format**, click **Proportions**. In the **Expected** proportions box, enter the range of cells containing the expected proportions. If you included column labels in the data ranges, click **Column labels**. Click **Chi-square test**. For the **Significance level (%)**, enter **5**. Click **OK**.

Chi-square test:	
Chi-square (Observed value)	1.7622
Chi-square (Critical value)	15.5073
DF	8
p-value	0.9875
alpha	0.05

Test statistic: $\chi^2 = 1.7622$. Critical value: $\chi^2 = 15.5073$. *P*-value = 0.9875. There is not sufficient evidence to warrant rejection of the claim that the leading digits are from a population with a distribution that conforms to Benford's law. The tax entries do appear to be legitimate.

Section 11-2, Beyond the Basics

25. a. 6, 13, 15, 6

  b. $z = \dfrac{155.41 - 162}{6.595} = -1; \ P(z < -1) = 0.1587 ;$

  $z = \dfrac{162.005 - 162}{6.595} = 0; \ P(-1 < z < 0) = 0.5000 - 0.1587 = 0.3413 ;$

  $z = \dfrac{168.601 - 162}{6.595} = 1; \ P(0 < z < 1) = 0.8413 - 0.5000 = 0.3413 :$

  $z = \dfrac{215 - 280}{65} = -1; \ P(z > 1) = 0.1587$

  c. $40 \cdot 0.1587 = 6.348$, $40 \cdot 0.3413 = 13.652$, $40 \cdot 0.3413 = 13.652$, $40 \cdot 0.1587 = 6.348$

  d. Test statistic: $\chi^2 = 0.202$. Critical value: $\chi^2 = 11.345$. *P*-value = 0.9772. There is not sufficient evidence to warrant rejection of the claim that heights were randomly selected from a normally distributed population. The test suggests that the data are from a normally distributed population.

Height	O	E	$O-E$	$(O-E)^2$	$\dfrac{(O-E)^2}{E}$
Less than 155.410	6	6.348	−0.348	0.121104	0.019078
155.410–162.005	13	13.652	−0.652	0.425104	0.031139
162.005–168.601	15	13.652	1.348	1.817104	0.133102
Greater than 168.601	6	6.348	−0.348	0.121104	0.019078
				Sum	0.202395

Section 11-3, Basic Skills and Concepts

1. Because the *P*-value of 0.216 is not small (such as 0.05 or lower), fail to reject the null hypothesis of independence between the treatment and whether the subject stops smoking. This suggests that the choice of treatment doesn't appear to make much of a difference.

3. $df = (3-1)(2-1) = 2$ and the critical value is $\chi^2 = 5.991$.

5. Enter the frequency counts of the contingency table in an Excel worksheet. Select the **XLSTAT** add-in. Select **Correlation/Association tests**. Select **Tests on contingency tables**. In the **Contingency table** box, enter the range of cells containing the frequency counts of the contingency table. Under Data format, select **Contingency table**. If you included labels, click **Labels included**. Click the **Options** tab. Click **Chi-square test**. Enter **5** for the **Significance level (%)**. Click **OK**. There is not sufficient evidence to warrant rejection of the claim that the form of the 100-Yuan gift is independent of whether the money was spent. There is not sufficient evidence to support the claim of a denomination effect.

Chi-square (Observed value)	3.4091
Chi-square (Critical value)	3.8415
DF	1
p-value	0.0648
alpha	0.05

7. Enter the frequency counts of the contingency table in an Excel worksheet. Select the **XLSTAT** add-in. Select **Correlation/Association tests**. Select **Tests on contingency tables**. In the **Contingency table** box, enter the range of cells containing the frequency counts of the contingency table. Under Data format, select **Contingency table**. If you included labels, click **Labels included**. Click the **Options** tab. Click **Chi-square test**. Enter **5** for the **Significance level (%)**. Click **OK**. There is sufficient evidence to warrant rejection of the claim that whether a subject lies is independent of the polygraph test indication. The results suggest that polygraphs are effective in distinguishing between truths and lies, but there are many false positives and false negatives, so they are not highly reliable.

Chi-square (Observed value)	25.5712
Chi-square (Critical value)	3.8415
DF	1
p-value	< 0.0001
alpha	0.05

9. Enter the frequency counts of the contingency table in an Excel worksheet. Select the **XLSTAT** add-in. Select **Correlation/Association tests**. Select **Tests on contingency tables**. In the **Contingency table** box, enter the range of cells containing the frequency counts of the contingency table. Under Data format, select **Contingency table**. If you included labels, click **Labels included**. Click the **Options** tab. Click **Chi-square test**. Enter **5** for the **Significance level (%)**. Click **OK**. There is sufficient evidence to warrant rejection of the claim that the sentence is independent of the plea. The results encourage pleas for guilty defendants.

Chi-square (Observed value)	42.5565
Chi-square (Critical value)	3.8415
DF	1
p-value	< 0.0001
alpha	0.05

11. Enter the frequency counts of the contingency table in an Excel worksheet. Select the **XLSTAT** add-in. Select **Correlation/Association tests**. Select **Tests on contingency tables**. In the **Contingency table** box, enter the range of cells containing the frequency counts of the contingency table. Under Data format, select **Contingency table**. If you included labels, click **Labels included**. Click the **Options** tab. Click **Chi-square test**. Enter **5** for the **Significance level (%)**. Click **OK**. There is not sufficient evidence to warrant rejection of the claim that the gender of the tennis player is independent of whether the call is overturned.

Chi-square (Observed value)	0.1636
Chi-square (Critical value)	3.8415
DF	1
p-value	0.6859
alpha	0.05

13. Enter the frequency counts of the contingency table in an Excel worksheet. Select the **XLSTAT** add-in. Select **Correlation/Association tests**. Select **Tests on contingency tables**. In the **Contingency table** box, enter the range of cells containing the frequency counts of the contingency table. Under Data format, select **Contingency table**. If you included labels, click **Labels included**. Click the **Options** tab. Click **Chi-square test**. Enter **5** for the **Significance level (%)**. Click **OK**. There is sufficient evidence to warrant rejection of the claim that the direction of the kick is independent of the direction of the goalkeeper jump. The results do not support the theory that because the kicks are so fast, goalkeepers have no time to react, so the directions of their jumps are independent of the directions of the kicks.

Chi-square (Observed value)	14.5887
Chi-square (Critical value)	9.4877
DF	4
p-value	0.0056
alpha	0.05

15. Enter the frequency counts of the contingency table in an Excel worksheet. Select the **XLSTAT** add-in. Select **Correlation/Association tests**. Select **Tests on contingency tables**. In the **Contingency table** box, enter the range of cells containing the frequency counts of the contingency table. Under Data format, select **Contingency table**. If you included labels, click **Labels included**. Click the **Options** tab. Click **Chi-square test**. Enter **5** for the **Significance level (%)**. Click **OK**. There is not sufficient evidence to warrant rejection of the claim that getting a cold is independent of the treatment group. The results suggest that echinacea is not effective for preventing colds.

Chi-square (Observed value)	2.9255
Chi-square (Critical value)	5.9915
DF	2
p-value	0.2316
alpha	0.05

17. Enter the frequency counts of the contingency table in an Excel worksheet. Select the **XLSTAT** add-in. Select **Correlation/Association tests**. Select **Tests on contingency tables**. In the **Contingency table** box, enter the range of cells containing the frequency counts of the contingency table. Under Data format, select **Contingency table**. If you included labels, click **Labels included**. Click the **Options** tab. Click **Chi-square test**. Enter **1** for the **Significance level (%)**. Click **OK**. There is sufficient evidence to warrant rejection of the claim that cooperation of the subject is independent of the age category. The age group of 60 and over appears to be particularly uncooperative.

Chi-square (Observed value)	20.2711
Chi-square (Critical value)	15.0863
DF	5
p-value	0.0011
alpha	0.01

19. Enter the frequency counts of the contingency table in an Excel worksheet. Select the **XLSTAT** add-in. Select **Correlation/Association tests**. Select **Tests on contingency tables**. In the **Contingency table** box, enter the range of cells containing the frequency counts of the contingency table. Under Data format, select **Contingency table**. If you included labels, click **Labels included**. Click the **Options** tab. Click **Chi-square test**. Enter 1 for the **Significance level (%)**. Click **OK**. There is not sufficient evidence to warrant rejection of the claim that getting an infection is independent of the treatment. The atorvastatin treatment does not appear to have an effect on infections.

Chi-square (Observed value)	0.7727
Chi-square (Critical value)	11.3449
DF	3
p-value	0.8560
alpha	0.01

Section 11-3, Beyond the Basics

21. Enter the frequency counts of the contingency table (Table 11.1) in an Excel worksheet. Select the **XLSTAT** add-in. Select **Correlation/Association tests**. Select **Tests on contingency tables**. In the **Contingency table** box, enter the range of cells containing the frequency counts of the contingency table. Under Data format, select **Contingency table**. If you included labels, click **Labels included**. Click the **Options** tab. Click **Chi-square test**. Enter 5 for the **Significance level (%)**. Click **OK**. Test statistics: $\chi^2 = 12.1619$ and $z = 3.4874$, so that $z^2 = \chi^2$. Critical values: $\chi^2 = 3.8415$ and $z^2 = \pm 1.96$, so $z^2 = \chi^2$ (approximately).

Chi-square (Observed value)	12.1619
Chi-square (Critical value)	3.8415
DF	1
p-value	0.0005
alpha	0.05

Chapter Quick Quiz

1. H_0: $p_1 = p_2 = p_3 = p_4 = p_5$. H_1: At least one of the probabilities is different from the others.

2. $O = 23$ and $E = \dfrac{107}{5} = 21.4$.

3. Right-tailed.

4. df = 4. Using Excel's CHISQ.INV.RT function, the critical value **=CHISQ.INV.RT(0.05,4)** = 9.4877.

5. There is not sufficient evidence to warrant rejection of the claim that occupation injuries occur with equal frequency on the different days of the week.

6. H_0: Response to the question is independent of gender. H_1: Response to the question and gender are dependent.

7. Chi-square distribution.

8. Right-tailed.

9. $df = (2-1)(3-1) = 2$. Using Excel's CHISQ.INV.RT function, the critical value **=CHISQ.INV.RT(0.05,2)** = 5.9915.

10. There is not sufficient evidence to warrant rejection of the claim that response is independent of gender.

Review Exercises

1. Enter the observed frequency data in an Excel worksheet. The expected proportion for each day of the week is 1/7 or 0.1429. Enter this expected proportion (or the formula =**1/7**) for each day of the week. Select the **XLSTAT** add-in. Select **Parametric tests**. Select **Multinomial good of fit test**. In the **Frequencies** box, enter the range of cells containing the observed frequencies. For **Data format**, click **Proportions**. In the **Expected** proportions box, enter the range of cells containing the expected proportions. If you included column labels in the data ranges, click **Column labels**. Click **Chi-square test**. For the **Significance level (%)**, enter **1**. Click **OK**.

Chi-square (Observed value)	931.3466
Chi-square (Critical value)	16.8119
DF	6
p-value	< 0.0001
alpha	0.01

Test statistic: $\chi^2 = 931.3466$. Critical value: $\chi^2 = 16.8119$. P-value: <0.0001. There is sufficient evidence to warrant rejection of the claim that auto fatalities occur on the different days of the week with the same frequency. Because people generally have more free time on weekends and more drinking occurs on weekends, the days of Friday, Saturday, and Sunday appear to have disproportionately more fatalities.

2. Enter the observed frequency data in an Excel worksheet. The expected proportion for each last digit is 1/10 or 0.1. Enter this expected proportion for each last digit. Select the **XLSTAT** add-in. Select **Parametric tests**. Select **Multinomial good of fit test**. In the **Frequencies** box, enter the range of cells containing the observed frequencies. For **Data format**, click **Proportions**. In the **Expected** proportions box, enter the range of cells containing the expected proportions. If you included column labels in the data ranges, click **Column labels**. Click **Chi-square test**. For the **Significance level (%)**, enter **5**. Click **OK**.

Chi-square test:	
Chi-square (Observed value)	6.5000
Chi-square (Critical value)	16.9190
DF	9
p-value	0.6890
alpha	0.05

Test statistic: $\chi^2 = 6.5000$. Critical value: $\chi^2 = 16.9190$. P-value = 0.6890. There is not sufficient evidence to warrant rejection of the claim that the last digits of 0, 1, 2, . . . , 9 occur with the same frequency. It does appear that the weights were obtained through measurements.

3. Enter the observed frequency data in an Excel worksheet. The expected proportion for each month is 1/12 or 0.0833. Enter this expected proportion (or the formula =**1/12**) for each month. Select the **XLSTAT** add-in. Select **Parametric tests**. Select **Multinomial good of fit test**. In the **Frequencies** box, enter the range of cells containing the observed frequencies. For **Data format**, click **Proportions**. In the **Expected** proportions box, enter the range of cells containing the expected proportions. If you included column labels in the data ranges, click **Column labels**. Click **Chi-square test**. For the **Significance level (%)**, enter **1**. Click **OK**.

Chi-square (Observed value)	288.4479
Chi-square (Critical value)	24.7250
DF	11
p-value	< 0.0001
alpha	0.01

Test statistic: $\chi^2 = 288.4479$. Critical value: $\chi^2 = 24.7250$. P-value < 0.0001. There is sufficient evidence to warrant rejection of the claim that weather-related deaths occur in the different months with the same frequency. The summer months appear to have disproportionately more weather-related deaths, and that is probably due to the fact that vacations and outdoor activities are much greater during those months.

4. Enter the frequency counts of the contingency table in an Excel worksheet. Select the **XLSTAT** add-in. Select **Correlation/Association tests**. Select **Tests on contingency tables**. In the **Contingency table** box, enter the range of cells containing the frequency counts of the contingency table. Under Data format, select **Contingency table**. If you included labels, click **Labels included**. Click the **Options** tab. Click **Chi-square test**. Enter **5** for the **Significance level (%)**. Click **OK**. There is sufficient evidence to warrant rejection of the claim that wearing a helmet has no effect on whether facial injuries are received. It does appear that a helmet is helpful in preventing facial injuries in a crash.

Chi-square (Observed value)	10.7081
Chi-square (Critical value)	3.8415
DF	1
p-value	0.0011
alpha	0.05

5. Enter the frequency counts of the contingency table in an Excel worksheet. Select the **XLSTAT** add-in. Select **Correlation/Association tests**. Select **Tests on contingency tables**. In the **Contingency table** box, enter the range of cells containing the frequency counts of the contingency table. Under Data format, select **Contingency table**. If you included labels, click **Labels included**. Click the **Options** tab. Click **Chi-square test**. Enter **5** for the **Significance level (%)**. Click **OK**. There is sufficient evidence to warrant rejection of the claim that when flipping or spinning a penny, the outcome is independent of whether the penny was flipped or spun. It appears that the outcome is affected by whether the penny is flipped or spun. If the significance level is changed to 0.01, the critical value changes to 6.6349, and we fail to reject the given claim, so the conclusion does change. All expected counts are greater than 5.

Chi-square (Observed value)	4.9554
Chi-square (Critical value)	3.8415
DF	1
p-value	0.0260
alpha	0.05

6. Enter the frequency counts of the contingency table in an Excel worksheet. Select the **XLSTAT** add-in. Select **Correlation/Association tests**. Select **Tests on contingency tables**. In the **Contingency table** box, enter the range of cells containing the frequency counts of the contingency table. Under Data format, select **Contingency table**. If you included labels, click **Labels included**. Click the **Options** tab. Click **Chi-square test**. Enter **5** for the **Significance level (%)**. Click **OK**. There is not sufficient evidence to warrant rejection of the claim that home/visitor wins are independent of the sport.

Chi-square (Observed value)	4.7372
Chi-square (Critical value)	7.8147
DF	3
p-value	0.1921
alpha	0.05

Cumulative Review Exercises

1. $H_0: p = 0.5.$ $H_1: p \neq 0.5$. Select the **XLSTAT** add-in. Select **Parametric tests**. Select **Tests for one proportion**. Frequency: **325**, Sample size: **489**, Test proportion: **0.5**, Data format: **Frequency**. Select **Sheet**. Select z-**test**. Click the **Options** tab. Alternative hypothesis: **Proportion – Test proportion ≠ D**, Hypothesized difference (D): **0**, Significance level (%): **5**. Click **OK**. Reject H_0. There is sufficient evidence to warrant rejection of the claim that among those who die in weather-related deaths, the percentage of males is equal to 50%.

Difference	0.1646
z (Observed value)	7.2807
z (Critical value)	1.9600
p-value (Two-tailed)	< 0.0001
alpha	0.05

2. Select the **XLSTAT** add-in. Select **Parametric tests**. Select **Tests for one proportion**. Proportion: **0.62**, Sample size: **1000**, Test proportion: **0.5**, Data format: **Proportion**. Select **Sheet**. Select z-**test**. Click the **Options** tab. Alternative hypothesis: **Proportion – Test proportion ≠ D**, Hypothesized difference (D): **0**, Significance level (%): **5**, Variance (confidence interval: **Sample**, Confidence interval: **Wald**. Click **OK**. $59.0\% < p < 65.0\%$. Because the confidence interval does not include 50% (or "half"), we should reject the stated claim.

95% confidence interval on the proportion (Wald):				
(0.5899, 0.6501)				

3. Enter the ages in an Excel worksheet. Select the **XLSTAT** add-in. Select **Describing data**. Select **Descriptive statistics**. In the **Quantitative data** box, enter the range of the age data. If you included a label, click **Sample labels**. Click the **Options** tab. Select **Descriptive statistics**. Click the **Outputs** tab. Select **Median, Mean**, and **Standard deviation (n-1)**. Click **OK**. Because an age of 16 differs from the mean by more than 2 standard deviations, it is an unusual age.

Statistic	Age
Median	60.0000
Mean	53.7000
Standard deviation (n-1)	16.0904

4. Enter the ages in an Excel worksheet. Select the **XLSTAT** add-in. Select **Describing data**. Select **Descriptive statistics**. In the **Quantitative data** box, enter the range of the age data. If you included a label, click **Sample labels**. Click the **Options** tab. Select **Descriptive statistics**. Click the **Outputs** tab. Enter **95** for the Confidence Interval (%). Click the **Outputs** tab. Select **Lower bound on mean (95%)** and **Upper bound on mean (95%)**. Click **OK**. The 95% confidence interval: 42.2 years $< \mu < 65.2$ years . Yes, the confidence interval limits do contain the value of 65.0 years that was found from a sample of 9269 ICU patients.

Statistic	Age
Lower bound on mean (95%)	42.1896
Upper bound on mean (95%)	65.2104

5. a. Enter the data in an Excel worksheet and be sure to include labels. Select the **XLSTAT** add-in. Select **Correlation/Association tests** data. Select **Correlation tests**. Enter the data range in the **Observations/variables table** box. For type of correlation, select **Pearson**. Select **Variable labels**. Set the Significance level (%) equal to **5**. Click the **Outputs** tab. Select **Correlations** and **p-values**. Click **OK**. $r = -0.0458$. Critical values: $r = \pm 0.632$. P-value $= 0.9000$. There is not sufficient evidence to support the claim that there is a linear correlation between the numbers of boats and the numbers of manatee deaths.

Correlation matrix (Pearson):		
Variables	Boats	Manatee Deaths
Boats	1	-0.0458
Manatee Deaths	-0.0458	1
p-values:		
Variables	Boats	Manatee Deaths
Boats	0	0.9000
Manatee Deaths	0.9000	0

  b. Enter the data in an Excel worksheet and be sure to include labels. Select the **XLSTAT** add-in. Select **Modeling data**. Select **Linear regression**. Enter the manatee deaths data range in the **Y/Dependent variables** box. Enter the number of boats data range in the **X/Explanatory variables** box. Click **Sheet**. Click **Variable labels**. Click **OK**. The equation of the regression line: $\hat{y} = 96.1 - 0.137x$

Equation of the model:		
Manatee Deaths = 96.14458-0.13655*Boats		

  c. $\hat{y} = 96.1 - 0.137(84) = 84.6$ manatee deaths (the value of y). The predicted value is not very accurate because it is not very close to the actual value of 78 manatee deaths.

6. a. Using Excel's NORM.INV function, the 5th percentile =**NORM.INV(0.05,686,34)** = 630 mm.

  b. Using Excel's NORM.DIST function, the percent of women who cannot reach the dashboard =**NORM.DIST(650,686,34,TRUE)** = 0.1448 or 14.48%. That percentage is too high, because too many women would not be accommodated.

c. $z = \dfrac{680 - 686}{34/\sqrt{16}} = -0.706$. $P(z > -0.706)$ **=1-NORM.S.DIST(-0.706,TRUE)** = 0.7599 or 75.99%.

Groups of 16 women do not occupy a cockpit; because individual women occupy the cockpit, this result has no effect on the design.

7.   a.   Statistic.

b.   Quantitative.

c.   Discrete.

d.   The sampling is conducted so that all samples of the same size have the same chance of being selected.

e.   The sample is a voluntary response sample (or self-selected sample), and those with strong feelings about the topic are more likely to respond, so it is not a valid sampling plan.

8.   a. $(0.6)^4 = 0.1296$

b. $1 - 0.6 = 0.4$

Chapter 12

Analysis of Variance

Section 12-2, Basic Skills and Concepts

1. a. The chest deceleration measurements are categorized according to the one characteristic of size.

   b. The terminology of analysis of variance refers to the method used to test for equality of the three population means. That method is based on two different estimates of a common population variance.

3. The test statistic is $F = 3.288$, and the F distribution applies.

5. Test statistic: $F = 0.39$. P-value: 0.677. Fail to reject H_0: $\mu_1 = \mu_2 = \mu_3$. There is not sufficient evidence to warrant rejection of the claim that the three categories of blood lead level have the same mean verbal IQ score. Exposure to lead does not appear to have an effect on verbal IQ scores.

7. Test statistic: $F = 11.6102$. P-value: 0.000577. Reject H_0: $\mu_1 = \mu_2 = \mu_3$. There is sufficient evidence to warrant rejection of the claim that the three size categories have the same mean highway fuel consumption. The size of a car does appear to affect highway fuel consumption.

9. Test statistic: $F = 0.161$. P-value: 0.852. Fail to reject H_0: $\mu_1 = \mu_2 = \mu_3$. There is not sufficient evidence to warrant rejection of the claim that the three size categories have the same mean head injury measurement. The size of a car does not appear to affect head injuries.

11. Enter the data in an Excel worksheet, placing the dependent variable values in one column and the corresponding category name for each dependent variable value in a separate column. The first three rows are shown on the right. Select the **XLSTAT** add-in. Select **Modeling data**. Select **ANOVA**. Enter the range of the dependent variable values in the **Y/Dependent variables** box. Enter the range of independent variable values (category names) in the **X/Explanatory variables** box. Be sure the **Qualitative** box is checked. If variable labels are included in the data ranges, check the **Variable labels** box. Click the **Outputs** tab. Select **Descriptive statistics** and **Analysis of Variance**. Click **OK**.

Time	Mile
3.15	1
3.24	1

Source	DF	Sum of squares	Mean squares	F	Pr > F
Model	2	0.0372	0.0186	27.2488	< 0.0001
Error	12	0.0082	0.0007		
Corrected Total	14	0.0454			

Test statistic: $F = 27.2488$. P-value: < 0.0001. Reject H_0: $\mu_1 = \mu_2 = \mu_3$. There is sufficient evidence to warrant rejection of the claim that the three different miles have the same mean time. These data suggest that the third mile appears to take longer, and a reasonable explanation is that the third lap has a hill.

13. Enter the data in an Excel worksheet, placing the dependent variable values in one column and the corresponding category name for each dependent variable value in a separate column. Select the **XLSTAT** add-in. Select **Modeling data**. Select **ANOVA**. Enter the range of the dependent variable values in the **Y/Dependent variables** box. Enter the range of independent variable values (category names) in the **X/Explanatory variables** box. Be sure the **Qualitative** box is checked. If variable labels are included in the data ranges, check the **Variable labels** box. Click the **Outputs** tab. Select **Descriptive statistics** and **Analysis of Variance**. Click **OK**.

Source	DF	Sum of squares	Mean squares	F	Pr > F
Model	3	3.3465	1.1155	6.1413	0.0056
Error	16	2.9062	0.1816		
Corrected Total	19	6.2527			

Test statistic: $F = 6.1413$. P-value: 0.0056. Reject H_0: $\mu_1 = \mu_2 = \mu_3 = \mu_4$. There is sufficient evidence to warrant rejection of the claim that the four treatment categories yield poplar trees with the same mean weight. Although not justified by the results from analysis of variance, the treatment of fertilizer and irrigation appears to be most effective.

15. Open the **CIGARET** data file on your data disk. Copy the data for the exercise to a new sheet and rearrange it so that the dependent variable values are in one column and the corresponding category name for each dependent variable value is in another column. Select the **XLSTAT** add-in. Select **Modeling data**. Select **ANOVA**. Enter the range of the dependent variable values in the **Y/Dependent variables** box. Enter the range of independent variable values (category names) in the **X/Explanatory variables** box. Be sure the **Qualitative** box is checked. If variable labels are included in the data ranges, check the **Variable labels** box. Click the **Outputs** tab. Select **Descriptive statistics** and **Analysis of Variance**. Click **OK**.

Source	DF	Sum of squares	Mean squares	F	Pr > F
Model	2	2.2083	1.1041	18.9931	< 0.0001
Error	72	4.1856	0.0581		
Corrected Total	74	6.3939			

Test statistic: $F = 18.9931$. P-value: < 0.0001. Reject H_0: $\mu_1 = \mu_2 = \mu_3$. There is sufficient evidence to warrant rejection of the claim that the three different types of cigarettes have the same mean amount of nicotine. Given that the king-size cigarettes have the largest mean of 1.26 mg per cigarette, compared to the other means of 0.87 mg per cigarette and 0.92 mg per cigarette, it appears that the filters do make a difference (although this conclusion is not justified by the results from analysis of variance).

Section 12-2, Beyond the Basics

17. The Tukey test results show different P-values, but they are not dramatically different. The Tukey results suggest the same conclusions as the Bonferroni test.

Section 12-3, Basic Skills and Concepts

1. The load values are categorized using two different factors of (1) femur (left or right) and (2) size of car (small, midsize, large).

3. An interaction between two factors or variables occurs if the effect of one of the factors changes for different categories of the other factor. If there is an interaction effect, we should not proceed with individual tests for effects from the row factor and column factor. If there is an interaction, we should not consider the effects of one factor without considering the effects of the other factor.

5. For interaction, the test statistic is $F = 1.72$ and the P-value is 0.194, so there is not sufficient evidence to conclude that there is an interaction effect. For the row variable of femur (right, left), the test statistic is $F = 1.39$ and the P-value is 0.246, so there is not sufficient evidence to conclude that whether the femur is right or left has an effect on measured load. For the column variable of size of the car, the test statistic is $F = 2.23$ and the P-value is 0.122, so there is not sufficient evidence to conclude that the car size category has an effect on the measured load.

7. For interaction, the test statistic is $F = 1.05$ and the P-value is 0.365, so there is not sufficient evidence to conclude that there is an interaction effect. For the row variable of sex, the test statistic is $F = 4.58$ and the P-value is 0.043, so there is sufficient evidence to conclude that the sex of the subject has an effect on verbal IQ score. For the column variable of blood lead level (LEAD), the test statistic is $F = 0.14$ and the P-value is 0.871, so there is not sufficient evidence to conclude that blood lead level has an effect on verbal IQ score. It appears that only the sex of the subject has an effect on verbal IQ score.

9. Enter the data in an Excel worksheet ensuring that the data are entered as shown in the middle of page 659 in the textbook. Click on the **DATA** tab in the Ribbon. Select **Data Analysis** in the Analysis group. (If Data Analysis is not available, you must install the *Analysis Toolpak* add-in as described in Section 1-5 of your textbook.) Select **Anova: Two-Factor With Replication**. Click **OK**. Click on the **Input Range** box, and enter the range containing the data for the exercise. In **Rows per sample**, enter **12** for the number of values in each cell. Set Alpha equal to **0.05**. Select **New Worksheet Ply**. Click **OK**.

ANOVA						
Source of Variation	SS	df	MS	F	P-value	F crit
Sample	4.5	1	4.5	4.977654	0.02908	3.986269
Columns	2.861111	2	1.430556	1.582402	0.213177	3.135918
Interaction	6.75	2	3.375	3.73324	0.029108	3.135918
Within	59.66667	66	0.90404			
Total	73.77778	71				

For interaction, the test statistic is $F = 3.7332$ and the P-value is 0.0291, so there is sufficient evidence to conclude that there is an interaction effect. The measures of self-esteem appear to be affected by an interaction between the self-esteem of the subject and the self-esteem of the target. Because there appears to be an interaction effect, we should not proceed with individual tests of the row factor (target's self-esteem) and the column factor (subject's self-esteem).

Section 12-3, Beyond the Basics

11. a.   Test statistics and P-values do not change.

    b.   Test statistics and P-values do not change.

    c.   Test statistics and P-values do not change.

    d.   An outlier can dramatically affect and change test statistics and P-values.

Chapter Quick Quiz

1. H_0: $\mu_1 = \mu_2 = \mu_3$. Because the displayed P-value of 0.000 is small, reject H_0.

2.   No. Because we reject the null hypothesis of equal means, it appears that the three different power sources do not produce the same mean voltage level, so we cannot expect electrical appliances to behave the same way when run from the three different power sources.

3.   Right-tailed.

4.   Test statistic: $F = 183.01$. In general, larger test statistics result in smaller P-values.

5.   The sample voltage measurements are categorized using only one factor: the source of the voltage.

6.   Test a null hypothesis that three or more samples are from populations with equal means.

7.   With one-way analysis of variance, the different samples are categorized using only one factor, but with two-way analysis of variance, the sample data are categorized into different cells determined by two different factors.

8.   For interaction, the test statistic is $F = 0.19$ and the P-value is 0.832. Fail to reject the null hypothesis of no interaction. There does not appear to be an effect due to an interaction between sex and major.

9.   The test statistic is $F = 0.78$ and the P-value is 0.395. There is not sufficient evidence to support a claim that the length estimates are affected by the sex of the subject.

10.   The test statistic is $F = 0.13$ and the P-value is 0.876. There is not sufficient evidence to support a claim that the length estimates are affected by the subject's major.

Review Exercises

1. H_0: $\mu_1 = \mu_2 = \mu_3$. Test statistic: $F = 10.10$. P-value: 0.001. Reject the null hypothesis. There is sufficient evidence to warrant rejection of the claim that 4-cylinder cars, 6-cylinder cars, and 8-cylinder cars have the same mean highway fuel consumption amount.

2.   For interaction, the test statistic is $F = 0.17$ and the P-value is 0.915, so there is not sufficient evidence to conclude that there is an interaction effect. For the row variable of site, the test statistic is $F = 0.81$ and the P-value is 0.374, so there is not sufficient evidence to conclude that the site has an effect on weight. For the column variable of treatment, the test statistic is $F = 7.50$ and the P-value is 0.001, so there is sufficient evidence to conclude that the treatment has an effect on weight.

3.   Open the **CIGARET** data file on your data disk. Copy the data for the exercise to a new sheet and rearrange it so that the dependent variable values are in one column and the corresponding category name for each dependent variable value is in another column. Select the **XLSTAT** add-in. Select **Modeling data**. Select **ANOVA**. Enter the range of the dependent variable values in the **Y/Dependent variables** box. Enter the range of independent variable values (category names) in the **X/Explanatory variables** box. Be sure the **Qualitative** box is checked. If variable labels are included in the data ranges, check the **Variable labels** box. Click the **Outputs** tab. Select **Descriptive statistics** and **Analysis of Variance**. Click **OK**.

Source	DF	Sum of squares	Mean squares	F	Pr > F
Model	2	1083.7067	541.8533	42.9436	< 0.0001
Error	72	908.4800	12.6178		
Corrected Total	74	1992.1867			

Test statistic: $F = 42.9436$. P-value: < 0.0001. Reject H_0: $\mu_1 = \mu_2 = \mu_3$. There is sufficient evidence to warrant rejection of the claim that the three different types of cigarettes have the same mean amount of tar. Given that the king-size cigarettes have the largest mean of 21.1 mg per cigarette, compared to the other means of 12.9 mg per cigarette and 13.2 mg per cigarette, it appears that the filters do make a difference (although this conclusion is not justified by the results from analysis of variance).

4.   Enter the data in an Excel worksheet ensuring that the data are entered as shown in the middle of page 659 in the textbook. Click on the **DATA** tab in the Ribbon. Select **Data Analysis** in the Analysis group. (If Data Analysis is not available, you must install the *Analysis Toolpak* add-in as described in Section 1-5 of your textbook. Select **Anova: Two-Factor With Replication**. Click **OK**. Click on the **Input Range** box, and enter the range containing the data for the exercise. In **Rows per sample**, enter **4** for the number of values in each cell. Set Alpha equal to **0.05**. Select **New Worksheet Ply**. Click **OK**.

Source of Variation	SS	df	MS	F	P-value	F crit
Sample	0.005625	1	0.005625	0.017822	0.896012	4.747225
Columns	0.950625	1	0.950625	3.011881	0.108238	4.747225
Interaction	0.275625	1	0.275625	0.873267	0.368476	4.747225
Within	3.7875	12	0.315625			
Total	5.019375	15				

For interaction, the test statistic is $F = 0.8733$ and the P-value is 0.3685, so there does not appear to be an effect from an interaction between gender and whether the subject smokes. For gender, the test statistic is $F = 0.0178$ and the P-value is 0.8960, so gender does not appear to have an effect on body temperature. For smoking, the test statistic is $F = 3.0119$ and the P-value is 0.1082, so there does not appear to be an effect from smoking on body temperature.

Cumulative Review Exercises

1.   Enter the data in an Excel worksheet, placing each group's data in a separate column. Be sure to include the labels. Select the **XLSTAT** add-in. Select **Describing data**. Select **Descriptive statistics**. In the **Quantitative data** box, enter the range of all three groups. Select **Sample labels**. Click the **Outputs** tab. Select **Mean**, **Variance (n-1)**, and **Standard deviation (n-1)**. Click **OK**.

Statistic	Presidents	Popes	Monarchs
Mean	15.5000	13.1250	22.7143
Variance (n-1)	94.5270	80.2880	346.0659
Standard deviation (n-1)	9.7225	8.9604	18.6028

a.   15.5 years, 13.1 years, 22.7 years

b.   9.7 years, 9.0 years, 18.6 years

c.   94.5 years2, 80.3 years^2, 346.1 years^2

d.   Ratio

2.   Enter the data in an Excel worksheet, placing each group's data in a separate column. Be sure to include the labels. Select the **XLSTAT** add-in. Select **Parametric tests**. Select **Two-sample *t*-test and *z*-test**. In the **Sample 1** box, enter the range of the presidents' data. In the **Sample 2** box, enter the range of the monarchs' data. Click **One column per sample**. Click **Column labels**. Select **Student's *t* test**. Click the **Options tab**. Alternative hypothesis: **Mean 1 − Mean 2 ≠ D**. Hypothesized difference (D): **0**. Significance level (%): **5**. Click **OK**. Fail to reject H_0: $\mu_1 = \mu_2$. There is not sufficient evidence to support the claim that there is a difference between the means for the two groups.

Difference	-7.2143
t (Observed value)	-1.3831
|t| (Critical value)	2.1233
DF	16
p-value (Two-tailed)	0.1860
alpha	0.05

3.   Enter the data in an Excel worksheet. Be sure to include the label. Select the **XLSTAT** add-in. Select **Describing data**. Select **Normality tests**. In the **Data** box, enter the range of the presidents' longevity data. Select **Sample labels**. Click the **Charts** tab. Select **Normal Q-Q plots**. Click **OK**. The data appear to come from a normal distribution, because the points in the normal quantile plot are reasonably close to a straight-line pattern with no other pattern that is not a straight-line pattern.

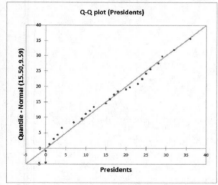

4.   Enter the data in an Excel worksheet. Be sure to include the label. Select the **XLSTAT** add-in. Select **Describing data**. Select **Descriptive statistics**. In the **Quantitative data** box, enter the range of the presidents' data. Click the **Options** tab. Select **Descriptive statistics**. For **Confidence interval (%)**, enter **95**. Click the **Outputs** tab. Select **Lower bound on mean (95%)** and **Upper bound on mean (95%)**. Click **OK**. 95% CI: 12.3 years $< \mu <$ 18.7 years .

Statistic	Presidents
Lower bound on mean (95%)	12.3043
Upper bound on mean (95%)	18.6957

5.   a. H_0: $\mu_1 = \mu_2 = \mu_3$

b.   Because the *P*-value of 0.051 is greater than the significance level of 0.05, fail to reject the null hypothesis of equal means. There is not sufficient evidence to warrant rejection of the claim that the three means are equal. The three populations do not appear to have means that are significantly different.

6.   Enter the data in an Excel worksheet, placing the September data and the April data in adjacent columns. Be sure to include the labels. Select the **XLSTAT** add-in. Select **Correlation/Association tests**. Select **Correlation tests**. In the **Observations/variables table** box, enter the range of the two variables. Type of correlation: **Pearson**. Select **Variable labels**. Significance level (%): **5**. Click the **Outputs** tab. Select **Correlations** and *p*-values. Click the **Charts** tab. Select **Scatter plots**. Click **OK**.

a. $r = 0.9181$. Critical values: $r = \pm 0.707$. P-value $= 0.0013$. There is sufficient evidence to support the claim that there is a linear correlation between September weights and the subsequent April weights.

Correlation matrix (Pearson):

Variables	September	April
September	1	0.9181
April	0.9181	1

p-values:

Variables	September	April
September	0	0.0013
April	0.0013	0

b.   Right-click on a plotted point in the scatter plot where September is on the x-axis and April is on the y-axis. Select **Add trendline**. Click **Linear**. Click **Display Equation on chart**. The equation of the regression line is $\hat{y} = 9.28 + 0.823x$.

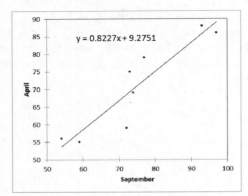

c. $\hat{y} = 9.28 + 0.823(94) = 86.6$ kg, which is not very close to the actual April weight of 105 kg.

7.   a. $z = \dfrac{345 - 280}{65} = 1;$ $P(z > 1) = 0.1587$.

b. $z = \dfrac{215 - 280}{65} = -1;$ $P(-1 < z < 1) = 0.8413 - 0.1587 = 0.6826$

c. $z = \dfrac{319 - 280}{65 / \sqrt{25}} = 3;$ $P(z < 3) = 0.9987$.

d.   80th percentile: $x = \mu + z \cdot \sigma = 280 + .84 \cdot 65 = 334.6$

8.   a. $0.20 \cdot 1000 = 200$ 200

b.   Select the **XLSTAT** add-in. Select **Parametric tests**. Select **Tests for one proportion**. Proportion: **0.20**. Sample size: **1000**. Test proportion: **0.50**. Data format: **Proportion**. Select z test. Click the **Options** tab. Alternative hypothesis: **Proportion – Test proportion $\neq$ D**. Hypothesized difference (D): **0**. Significance level (%): **5**. Variance (confidence interval): **Sample**. Confidence interval: **Wald**. Click **OK**. The 95% CI: $0.175 < p < 0.225$.

95% confidence interval on the proportion (Wald):

(0.1752, 0.2248)			

c.   Yes. The confidence interval shows us that we have 95% confidence that the true population proportion is contained within the limits of 0.175 and 0.225, and 1/4 is not included within that range.

9. a. The distribution should be uniform, with a flat shape. The given histogram agrees (approximately) with the uniform distribution that we expect.

    b. No. A normal distribution is approximately bell-shaped, but the given histogram is far from being bell-shaped.

10. Enter the observed frequency data in an Excel worksheet. The expected proportion for each digit is 1/10 or 0.1. Enter this expected proportion for each digit. Select the **XLSTAT** add-in. Select **Parametric tests**. Select **Multinomial good of fit test**. In the **Frequencies** box, enter the range of cells containing the observed frequencies. For **Data format**, click **Proportions**. In the **Expected** proportions box, enter the range of cells containing the expected proportions. If you included column labels in the data ranges, click **Column labels**. Click **Chi-square test**. For the **Significance level (%)**, enter **5**. Click **OK**.

Chi-square (Observed value)	10.4000
Chi-square (Critical value)	16.9190
DF	9
p-value	0.3191
alpha	0.05

There is not sufficient evidence to warrant rejection of the claim that the digits are selected from a population in which the digits are all equally likely. There does not appear to be a problem with the lottery.

Chapter 13

Nonparametric Statistics

Section 13-2, Basic Skills and Concepts

1. The only requirement for the matched pairs is that they constitute a simple random sample. There is no requirement of a normal distribution or any other specific distribution. The sign test is "distribution free" in the sense that it does not require a normal distribution or any other specific distribution.

3. H_0: There is no difference between the populations of September weights and April weights. H_1: There is a difference between the populations of September weights and April weights. The sample data do not contradict H_1 because the numbers of positive signs (2) and negative signs (7) are not exactly the same.

5. Using Excel's BINOM.DIST function, the two-tailed P-value $=$**2*BINOM.DIST(1,14,0.5,TRUE)** $= 0.0018$. Also, the test statistic of $x=1$ is less than or equal to the critical value of 2 (from Table A-7.) There is sufficient evidence to warrant rejection of the claim of no difference. There does appear to be a difference.

7. Using Excel's BINOM.DIST function, the two-tailed P-value $=$**2*BINOM.DIST(151,1007,0.5,TRUE)** $=$ 4.1871E-120. The test statistic of $z = \dfrac{(151+0.5) - \frac{1007}{2}}{\sqrt{1007}/2} = -22.18$ falls in the critical region bounded by

   $z=-1.96$ and 1.96. There is sufficient evidence to warrant rejection of the claim of no difference. There does appear to be a difference.

9. Open the **OSCR** data file on your data disk. Select the **XLSTAT** add-in. Select **Nonparametrics tests**. Select **Comparison of two samples**. In the **Sample 1** box, enter the data range for actresses, **A1:A11**. In the **Sample 2** box, enter the data range for actors, **B1:B11**. Under Data format, check **Paired samples**. Select **Column labels**. Select **Sign test**. Click the **Options** tab. Alternative hypothesis: **Sample 1 – Sample 2 ≠ D**. Hypothesized difference (D): **0**. Significance level (%): **5**. Select **Exact p-value**. Click **OK**. The two-tailed P-value $= 0.0215$. There is sufficient evidence to warrant rejection of the claim of no difference. There does appear to be a difference.

Sign test / Two-tailed test:	
N+	1
Expected value	5.0000
Variance (N+)	2.5000
p-value (Two-tailed)	0.0215
alpha	0.05

11. Open the **POTUS** data file on your data disk. Select the **XLSTAT** add-in. Select **Nonparametrics tests**. Select **Comparison of two samples**. In the **Sample 1** box, enter the data range for Ht, **E1:E39**. In the **Sample 2** box, enter the data range for HtOpp, **F1:F39**. Under Data format, check **Paired samples**. Select **Column labels**. Select **Sign test**. Click the **Options** tab. Alternative hypothesis: **Sample 1 – Sample 2 ≠ D**. Hypothesized difference (D): **0**. Significance level (%): **5**. Select **Exact p-value**. Click the **Missing data** tab. Select **Remove the observations**. Click **OK**. The two-tailed P-value $= 0.5966$. There is not sufficient evidence to warrant rejection of the claim of no difference. There does not appear to be a difference.

Sign test / Two-tailed test:	
N+	18
Expected value	16.0000
Variance (N+)	8.0000
p-value (Two-tailed)	0.5966
alpha	0.05

13. The test statistic of $z = \dfrac{(52+0.5) - \frac{291}{2}}{\sqrt{291}/2} = -10.90$ is in the critical region bounded by $z=\pm2.575$. Using

   Excel's BINOM.DIST function, the two-tailed P-value $=$**2*BINOM.DIST(52,291,0.5,TRUE)** $= 8.22462E-30$. There is sufficient evidence to warrant rejection of the claim of no difference. The YSORT method appears to have an effect on the gender of the child. (Because so many more boys were born than would be expected with no effect, it appears that the YSORT method is effective in increasing the likelihood that a baby will be a boy.)

15. The test statistic of $z = \frac{(123+0.5)-\frac{280}{2}}{\sqrt{280}/2} = -1.97$ is not in the critical region bounded by $z = \pm 2.575$. Using Excel's BINOM.DIST function, the two-tailed P-value =2*BINOM.DIST(123,280,0.5,TRUE) = 0.0484. There is not sufficient evidence to warrant rejection of the claim that the touch therapists make their selections with a method equivalent to random guesses. The touch therapists do not appear to be effective in selecting the correct hand.

17. Open the **COINS** data file on your data disk. Enter a column of 40 values all equal to **5.670**. Select the **XLSTAT** add-in. Select **Nonparametrics tests**. Select **Comparison of two samples**. Enter the data ranges in the Sample boxes. Under Data format, check **Paired samples**. If you included column labels in the data ranges, select **Column labels**. Select **Sign test**. Click the **Options** tab. Alternative hypothesis: **Sample 1 – Sample 2 ≠ D**. Hypothesized difference (D): **0**. Significance level (%): **1**. Select **Exact p-value**. Click **OK**.

N+	12
Expected value	20.0000
Variance (N+)	10.0000
p-value (Two-tailed)	0.0166
alpha	0.01

The test statistic of $z = \frac{(12+0.5)-\frac{40}{2}}{\sqrt{40}/2} = -2.37$ is not in the critical region bounded by $z = \pm 2.575$. There is not sufficient evidence to warrant rejection of the claim that the median is equal to 5.670 g. The quarters appear to be minted according to specifications.

19. Open the **COLA** data file on your data disk. Enter a column of 36 values all equal to **12**. Select the **XLSTAT** add-in. Select **Nonparametrics tests**. Select **Comparison of two samples**. Enter the data ranges in the Sample boxes. Under Data format, check **Paired samples**. If you included column labels in the data ranges, select **Column labels**. Select **Sign test**. Click the **Options** tab. Alternative hypothesis: **Sample 1 – Sample 2 ≠ D**. Hypothesized difference (D): **0**. Significance level (%): **5**. Select **Exact p-value**. Click **OK**.

N+	33
Expected value	17.0000
Variance (N+)	8.5000
p-value (Two-tailed)	< 0.0001
alpha	0.05

The test statistic of $z = \frac{(1+0.5)-\frac{34}{2}}{\sqrt{34}/2} = -5.32$ is in the critical region bounded by $z = \pm 1.96$. There is sufficient evidence to warrant rejection of the claim that the median amount of Coke is equal to 12 oz. Consumers are not being cheated because they are generally getting more than 12 oz of Coke, not less.

Section 13-2, Beyond the Basics

21. Second approach: The test statistic of $z = \frac{(30+0.5)-\frac{105}{2}}{\sqrt{105}/2} = -4.29$ is in the critical region bounded by $z = -1.645$, so the conclusions are the same as in Example 4.

Third approach: The test statistic of $z = \frac{(38+0.5)-\frac{106}{2}}{\sqrt{106}/2} = -2.82$ is in the critical region bounded by $z = -1.645$, so the conclusions are the same as in Example 4. The different approaches can lead to very different results; as seen in the test statistics of –4.21, –4.29, and –2.82. The conclusions are the same in this case, but they could be different in other cases.

Section 13-3, Basic Skills and Concepts

1. The only requirements are that the matched pairs be a simple random sample and the population of differences be approximately symmetric. There is no requirement of a normal distribution or any other specific distribution. The Wilcoxon signed-ranks test is "distribution free" in the sense that it does not require a normal distribution or any other specific distribution.

3. The sign test uses only the signs of the differences, but the Wilcoxon signed-ranks test uses ranks that are affected by the magnitudes of the differences.

5. Open the **OSCR** data file on your data disk. Select the **XLSTAT** add-in. Select **Nonparametrics tests**. Select **Comparison of two samples**. In the **Sample 1** box, enter the data range for actresses, **A1:A11**. In the **Sample 2** box, enter the data range for actors, **B1:B11**. Under Data format, check **Paired samples**. Select **Column labels**. Select **Wilcoxon signed-rank test**. Click the **Options** tab. Alternative hypothesis: **Sample 1 – Sample 2 ≠ D**. Significance level (%): **5**. Select **Exact p-value**. Click **OK**. Two-tailed *P*-value = 0.0322. Reject the null hypothesis that the population of differences has a median of 0. There is sufficient evidence to warrant rejection of the claim of no difference. There does appear to be a difference.

V	6.0000
Expected value	27.5000
Variance (V)	96.1250
p-value (Two-tailed)	0.0322

7. Open the **POTUS** data file on your data disk. Select the **XLSTAT** add-in. Select **Nonparametrics tests**. Select **Comparison of two samples**. In the **Sample 1** box, enter the data range for Ht, **E1:E39**. In the **Sample 2** box, enter the data range for HtOpp, **F1:F39**. Under Data format, check **Paired samples**. Select **Column labels**. Select **Wilcoxon signed-rank test**. Click the **Options** tab. Alternative hypothesis: **Sample 1 – Sample 2 ≠ D**. Hypothesized difference (D): **0**. Significance level (%): **5**. Select **Exact p-value**. Click the **Missing data** tab. Select **Remove the observations**. Click **OK**. The two-tailed *P*-value = 0.7573. Fail to reject the null hypothesis that the population of differences has a median of 0. There is not sufficient evidence to warrant rejection of the claim of no difference. There does not appear to be a difference.

V	281.0000
Expected value	264.0000
Variance (V)	2851.7500
p-value (Two-tailed)	0.7573
alpha	0.05

9. Open the **COINS** data file on your data disk. Enter a column of 40 values all equal to **5.670**. Select the **XLSTAT** add-in. Select **Nonparametrics tests**. Select **Comparison of two samples**. Enter the data ranges in the Sample boxes. Under Data format, check **Paired samples**. If you included column labels in the data ranges, select **Column labels**. Select **Wilcoxon signed-rank test**. Click the **Options** tab. Alternative hypothesis: **Sample 1 – Sample 2 ≠ D**. Hypothesized difference (D): **0**. Significance level (%): **1**. Select **Exact p-value**. Click **OK**. The two-tailed *P*-value = 0.0034. There is sufficient evidence to warrant rejection of the claim that the median is equal to 5.670 g. The quarters do not appear to be minted according to specifications.

V	196
Expected value	410.0000
Variance (V)	5535.0000
p-value (Two-tailed)	0.0034
alpha	0.01

11. Open the **COLA** data file on your data disk. Enter a column of 36 values all equal to **12**. Select the **XLSTAT** add-in. Select **Nonparametrics tests**. Select **Comparison of two samples**. Enter the data ranges in the Sample boxes. Under Data format, check **Paired samples**. If you included column labels in the data ranges, select **Column labels**. Select **Sign test**. Click the **Options** tab. Alternative hypothesis: **Sample 1 – Sample 2 ≠ D**. Hypothesized difference (D): **0**. Significance level (%): **5**. Select **Exact p-value**. Click **OK**. The two-tailed *P*-value < 0.0001. There is sufficient evidence to warrant rejection of the claim that the median amount of Coke is equal to 12 oz. Consumers are not being cheated because they are generally getting more than 12 oz of Coke, not less.

V	0.0000
Expected value	333.0000
Variance (V)	4050.7500
p-value (Two-tailed)	< 0.0001
alpha	0.05

Section 13-3, Beyond the Basics

13. a. Min: 0 and Max: $1 + 2 + \ldots + 74 + 75 = 2850$

b. $\dfrac{2850}{2} = 1425$

c. $2850 - 850 = 2000$

d. $\dfrac{n(n+1)}{2} - k$

Section 13-4, Basic Skills and Concepts

1. Yes. The two samples are independent because the flight data are not matched. The samples are simple random samples. Each sample has more than 10 values.

3. H_0: Arrival delay times from Flights 19 and 21 have the same median. There are three different possible alternative hypotheses: H_1: Arrival delay times from Flights 19 and 21 have different medians. H_1: Arrival delay times from Flight 19 have a median greater than the median of arrival delay times from Flight 21. H_1: Arrival delay times from Flight 19 have a median less than the median of arrival delay times from Flight 21.

5. Enter the data in an Excel worksheet. Select the **XLSTAT** add-in. Select **Nonparametrics tests**. Select **Comparison of two samples**. Enter the data ranges in the Sample boxes. Under Data format, check **One column per sample**. Select **Column labels**. Select **Mann-Whitney test**. Click the **Options** tab. Alternative hypothesis: **Sample 1 – Sample 2 ≠ D**. Hypothesized difference (D): **0**. Significance level (%): **5**. Select **Asymptotic p-value**. Click **OK**. The two-tailed P-value = 0.4879. Fail to reject the null hypothesis that the populations have the same median. There is not sufficient evidence to warrant rejection of the claim that Flights 19 and 21 have the same median arrival delay time.

U	59.5000
Expected value	72.0000
Variance (U)	299.3478
p-value (Two-tailed)	0.4879
alpha	0.05

7. Enter the data in an Excel worksheet. Select the **XLSTAT** add-in. Select **Nonparametrics tests**. Select **Comparison of two samples**. Enter the data ranges in the Sample boxes. Under Data format, check **One column per sample**. Select **Column labels**. Select **Mann-Whitney test**. Click the **Options** tab. Alternative hypothesis: **Sample 1 – Sample 2 ≠ D**. Hypothesized difference (D): **0**. Significance level (%): **5**. Select **Asymptotic p-value**. Click **OK**. The two-tailed P-value = 0.0006. Reject the null hypothesis that the populations have the same median. There is sufficient evidence to reject the claim that for those treated with 20 mg of atorvastatin and those treated with 80 mg of atorvastatin, changes in LDL cholesterol have the same median. It appears that the dosage amount does have an effect on the change in LDL

U	162.5000
Expected value	91.0000
Variance (U)	422.4630
p-value (Two-tailed)	0.0006
alpha	0.05

9. Open the **IQLEAD** data file on your data disk. Select the **XLSTAT** add-in. Select **Nonparametrics tests**. Select **Comparison of two samples**. Enter the data range of the medium lead level group (group 2), H80:H101, in the Sample 1 box. Enter the data range of the high lead level group (group 3), H102:H122), in the Sample 2 box. Under Data format, check **One column per sample**. Do not select **Column labels**. Select **Mann-Whitney test**. Click the **Options** tab. Alternative hypothesis: **Sample 1 – Sample 2 > D**. Hypothesized difference (D): **0**. Significance level (%): **5**. Select **Asymptotic p-value**. Click **OK**. The two-tailed P-value = 0.3441. Fail to reject the null hypothesis that the populations have the same median. There is not sufficient evidence to support the claim that subjects with medium lead levels have full IQ scores with a higher median than the median full IQ score for subjects with high lead levels. It does not appear that lead level affects full IQ scores.

U	248.0000
Expected value	231.0000
Variance (U)	1691.0581
p-value (one-tailed)	0.3441
alpha	0.05

11. Open the COINS data file on your data disk. Select the **XLSTAT** add-in. Select **Nonparametrics tests**. Select **Comparison of two samples**. Enter the data ranges in the Sample boxes. Under Data format, check **One column per sample**. Select **Column labels**. Select **Mann-Whitney test**. Click the **Options** tab. Alternative hypothesis: **Sample 1 – Sample 2 ≠ D**. Hypothesized difference (D): **0**. Significance level (%): **5**. Select **Asymptotic p-value**. Click **OK**. The two-tailed P-value < 0.0001. Reject the null hypothesis that the populations have the same median. It appears that the design of quarters changed in 1964.

U	1600.0000
Expected value	800.0000
Variance (U)	10800.0000
p-value (Two-tailed)	< 0.0001
alpha	0.05

Section 13-4, Beyond the Basics

13. Using $U = 12 \cdot 11 + \dfrac{12(12+1)}{2} - 123.5 = 86.5$, we get $z = \dfrac{86.5 - \dfrac{12 \cdot 11}{2}}{\sqrt{\dfrac{12 \cdot 11 \cdot (12 + 11 + 1)}{12}}} = 1.26$. The test statistic is the same value with opposite sign.

Section 13-5, Basic Skills and Concepts

1. $R_1 = 1 + 10 + 12.5 + 5 + 8 = 36.5$, $R_2 = 3 + 14 + 15 + 2 + 12.5 + 6 = 52.5$, $R_3 = 7 + 16 + 10 + 4 + 10 = 47$

Low Lead Level	Medium Lead Level	High Lead Level
70 (1)	72 (3)	82 (7)
85 (10)	90 (14)	93 (16)
86 (12.5)	92 (15)	85 (10)
76 (5)	71 (2)	75 (4)
84 (8)	86 (12.5)	85 (10)
	79 (6)	

3. $n_1 = 5$, $n_2 = 6$, $n_3 = 5$, and $N = 5 + 6 + 5 = 16$.

5. Enter the data in an Excel worksheet. Select the **XLSTAT** add-in. Select **Nonparametrics tests**. Select **Comparison of K samples**. Enter the data range in the **Samples** box. Under Data format, check **One column**

per sample. Select **Column labels**. Select **Kruskal-Wallis test**. Click the **Options** tab. Significance level (%): **5**. Select **Asymptotic p-value**. Click **OK**. The P-value = 0.0065. Reject the null hypothesis of equal medians. The data suggest that the different miles present different levels of difficulty.

K (Observed value)	10.0699
K (Critical value)	5.9915
DF	2
p-value (Two-tailed)	0.0065
alpha	0.05

7. Enter the data in an Excel worksheet. Select the **XLSTAT** add-in. Select **Nonparametrics tests**. Select **Comparison of K samples**. Enter the data range in the **Samples** box. Under Data format, check **One column per sample**. Select **Column labels**. Select **Kruskal-Wallis test**. Click the **Options** tab. Significance level (%): **5**. Select **Asymptotic p-value**. Click **OK**. The P-value = 0.0852. Fail to reject the null hypothesis of equal medians. The data do not suggest that larger cars are safer.

K (Observed value)	4.9246
K (Critical value)	5.9915
DF	2
p-value (Two-tailed)	0.0852
alpha	0.05

9. Open the **IQLEAD** data file on your data disk. Select the **XLSTAT** add-in. Select **Nonparametrics tests**. Select **Comparison of K samples**. Enter the **IQP** data range in the **Data** box. Enter the **LEAD** data range in the **Sample identifiers** box. Under Data format, check **One column per variable**. Select **Column labels**. Select **Kruskal-Wallis test**. Click the **Options** tab. Significance level (%): **1**. Select **Asymptotic p-value**. Click **OK**. The P-value = 0.0181. Fail to reject the null hypothesis of equal medians. The data do not suggest that lead exposure has an adverse effect.

K (Observed value)	8.0236
K (Critical value)	9.2103
DF	2
p-value (Two-tailed)	0.0181
alpha	0.01

11. Open the **CIGARET** data file on your data disk. Copy the KgNic, MN Nic, and FLNic data and place the data in adjacent columns of the Excel worksheet. Select the **XLSTAT** add-in. Select **Nonparametrics tests**. Select **Comparison of K samples**. Enter data range in the **Samples** box. Under Data format, check **One column per sample**. Select **Column labels**. Select **Kruskal-Wallis test**. Click the **Options** tab. Significance level (%): **5**. Select **Asymptotic p-value**. Click **OK**. The P-value < 0.0001. Reject the null hypothesis of equal medians. There is sufficient evidence to warrant rejection of the claim that the three different types of cigarettes have the same median amount of nicotine. It appears that the filters do make a difference.

K (Observed value)	29.0701
K (Critical value)	5.9915
DF	2
p-value (Two-tailed)	< 0.0001
alpha	0.05

Section 13-5, Beyond the Basics

13. Using $\Sigma T = 16,836$ (see table on the next page) and $N = 25 + 25 + 25 = 75$, the corrected value of H is

$$\frac{27.9098}{1 - \dfrac{16,836}{75^3 - 75}} = 29.0701,$$ which is not substantially different from the value found in Exercise 11.

In this case, the large numbers of ties do not appear to have a dramatic effect on the test statistic H.

Nicotine Level	Rank	t	$t^3 - t$
0.2	1.5	2	6
0.6	4.5	2	6
0.7	6.5	2	6
0.8	17.0	19	6840
0.9	28.0	3	24
1.0	33.5	8	504
1.1	48.0	21	9240
1.2	61.0	5	120
1.3	65.5	4	60
1.4	69.0	3	24
1.7	73.5	2	6
	SUM		16,836

Section 13-6, Basic Skills and Concepts

1. The methods of Section 10-3 should not be used for predictions. The regression equation is based on a linear correlation between the two variables, but the methods of this section do not require a linear relationship. The methods of this section could suggest that there is a correlation with paired data associated by some nonlinear relationship, so the regression equation would not be a suitable model for making predictions.

3. r represents the linear correlation coefficient computed from sample paired data; ρ represents the parameter of the linear correlation coefficient computed from a population of paired data; r_s denotes the rank correlation coefficient computed from sample paired data; ρ_s represents the rank correlation coefficient computed from a population of paired data. The subscript s is used so that the rank correlation coefficient can be distinguished from the linear correlation coefficient r. The subscript does not represent the standard deviation s. It is used in recognition of Charles Spearman, who introduced the rank correlation method.

5. $r_s = 1$. Critical values are $r_s = \pm 0.886$ (From Table A-9.) Reject the null hypothesis of $\rho_s = 0$. There is sufficient evidence to support a claim of a correlation between distance and time.

7. Enter the data in an Excel worksheet. Select the **XLSTAT** add-in. Select **Correlation/Association tests**. Select **Correlation tests**. In the **Observations/variables table** box, enter the data range. Type of Correlation: **Spearman**. Click **Variable labels**. Significance level (%): **5**. Click **OK**. Reject the null hypothesis of $\rho_s = 0$. There is sufficient evidence to support the claim of a correlation between the quality scores and prices. These results do suggest that you get better quality by spending more.

Correlation matrix (Spearman):

Variables	Quality	Price
Quality	1	0.8214
Price	0.8214	1

p-values:

Variables	Quality	Price
Quality	0	0.0341
Price	0.0341	0

9. Enter the data in an Excel worksheet. Select the **XLSTAT** add-in. Select **Correlation/Association tests**. Select **Correlation tests**. In the **Observations/variables table** box, enter the data range. Type of Correlation: **Spearman**. Click **Variable labels**. Significance level (%): **5**. Click **OK**. Reject the null hypothesis of $\rho_s = 0$. There is sufficient evidence to support the claim of a correlation between the two judges. Examination of the results shows that the first and second judges appear to have opposite rankings.

Correlation matrix (Spearman):

Variables	First	Second
First	1	-0.9286
Second	-0.9286	1

p-values:

Variables	First	Second
First	0	0.0067
Second	0.0067	0

11. Enter the data in an Excel worksheet. Select the **XLSTAT** add-in. Select **Correlation/Association tests**. Select **Correlation tests**. In the **Observations/variables table** box, enter the data range. Type of Correlation: **Spearman**. Click **Variable labels**. Significance level (%): **5**. Click **OK**. Reject the null hypothesis of $\rho_s = 0$. There is sufficient evidence to conclude that there is a correlation between overhead widths of seals from photographs and the weights of the seals.

Correlation matrix (Spearman):

Variables	Overhead	Weight
Overhead	1	1.0000
Weight	1.0000	1

p-values:

Variables	Overhead	Weight
Overhead	0	< 0.0001
Weight	< 0.0001	0

13. Open the **MBODY** data file on your data disk. Select the **XLSTAT** add-in. Select **Correlation/Association tests**. Select **Correlation tests**. In the **Observations/variables table** box, enter the data range. Type of Correlation: **Spearman**. Click **Variable labels**. Significance level (%): **5**. Click **OK**. Reject the null hypothesis of $\rho_s = 0$. There is sufficient evidence to conclude that there is a correlation between the systolic and diastolic blood pressure levels in males.

Correlation matrix (Spearman):

Variables	SYS	DIAS
SYS	1	0.3937
DIAS	0.3937	1

p-values:

Variables	SYS	DIAS
SYS	0	0.0124
DIAS	0.0124	0

15. Open the **FLIGHTS** data file on your data disk. Copy the Dep Delay column and paste it in column F adjacent to Arr Delay. Select the **XLSTAT** add-in. Select **Correlation/Association tests**. Select **Correlation tests**. In the **Observations/variables table** box, enter the data range. Type of Correlation: **Spearman**. Click **Variable labels**. Significance level (%): **5**. Click **OK**. Reject the null hypothesis of $\rho_s = 0$. There is sufficient evidence to conclude that there is a correlation between departure delay times and arrival delay times.

Correlation matrix (Spearman):

Variables	Arr Delay	Dep Delay
Arr Delay	1	0.6508
Dep Delay	0.6508	1

p-values:

Variables	Arr Delay	Dep Delay
Arr Delay	0	< 0.0001
Dep Delay	< 0.0001	0

Section 13-6, Beyond the Basics

17. a. $r_s = \pm\sqrt{\dfrac{2.447^2}{2.447^2 + 8 - 2}} = \pm 0.707$ is not very close to the values of $r_s = \pm 0.738$ found in Table A-9.

   b. $r_s = \pm\sqrt{\dfrac{2.763^2}{2.763^2 + 30 - 2}} = \pm 0.463$ is quite close to the values of $r_s = \pm 0.467$ found in Table A-9.

Section 13-7, Basic Skills and Concepts

1. No. The runs test can be used to determine whether the sequence of World Series wins by American League teams and National League teams is not random, but the runs test does not show whether the proportion of wins by the American League is significantly greater than 0.5.

3. a. Answers vary, but here is a sequence that leads to rejection of randomness because the number of runs is 2, which is very low: W W W W W W W W W W W W W E E E E E E E E

   b. Answers vary, but here is a sequence that leads to rejection of randomness because the number of runs is 17, which is very high: W E W E W E W E W E W E W E W E W W W W

5. $n_1 = 19$, $n_2 = 15$, $G = 16$, critical values: 11, 24 (From Table A-10.) Fail to reject randomness. There is not sufficient evidence to support the claim that we elect Democrats and Republicans in a sequence that is not random. Randomness seems plausible here.

7. $n_1 = 20$, $n_2 = 10$, $G = 16$, critical values: 9, 20 (From Table A-10.) Fail to reject randomness. There is not sufficient evidence to reject the claim that the dates before and after July 1 are randomly selected.

9. $n_1 = 24$, $n_2 = 21$, $G = 17$, $\mu_G = \dfrac{2 \cdot 24 \cdot 21}{24 + 21} + 1 = 23.4$, $\sigma_G = \sqrt{\dfrac{2 \cdot 24 \cdot 21 (2 \cdot 24 \cdot 21 - 24 - 21)}{(24 + 21)^2 (24 + 21 - 1)}} = 3.3007$. Test

   statistic: $z = \dfrac{17 - 23.4}{3.3007} = -1.94$. Critical values: $z = \pm 1.96$. (Tech: P-value = 0.05252.) Fail to reject randomness. There is not sufficient evidence to reject randomness. The runs test does not test for disproportionately more occurrences of one of the two categories, so the runs test does not suggest that either conference is superior.

11. The median is 2453, $n_1 = 23$, $n_2 = 23$, $G = 4$, $\mu_G = \dfrac{2 \cdot 23 \cdot 23}{23 + 23} + 1 = 24$,

   $\sigma_G = \sqrt{\dfrac{2 \cdot 23 \cdot 23 (2 \cdot 23 \cdot 23 - 23 - 23)}{(23 + 23)^2 (23 + 23 - 1)}} = 3.3553$. Test statistic: $z = \dfrac{4 - 24}{3.3553} = -5.96$. Critical values: $z = \pm 1.96$.

   (Tech: P-value = 0.0000.) Reject randomness. The sequence does not appear to be random when considering values above and below the median. There appears to be an upward trend, so the stock market appears to be a profitable investment for the long term, but it has been more volatile in recent years.

Section 13-7, Beyond the Basics

13. a.   No solution provided.

b.   The 84 sequences yield these results: 2 sequences have 2 runs, 7 sequences have 3 runs, 20 sequences have 4 runs, 25 sequences have 5 runs, 20 sequences have 6 runs, and 10 sequences have 7 runs.

c.   With $P(2 \text{ runs}) = 2/84$, $P(3 \text{ runs}) = 7/84$, $P(4 \text{ runs}) = 20/84$, $P(5 \text{ runs}) = 25/84$, $P(6 \text{ runs}) = 20/84$, and $P(7 \text{ runs}) = 10/84$, each of the G values of 3, 4, 5, 6, 7 can easily occur by chance, whereas $G = 2$ is unlikely because $P(2 \text{ runs})$ is less than 0.025. The lower critical value of G is therefore 2, and there is no upper critical value that can be equaled or exceeded.

d.   Critical value of $G = 2$ agrees with Table A-10. The table lists 8 as the upper critical value, but it is impossible to get 8 runs using the given elements.

Chapter Quick Quiz

1.   Distribution-free test

2.   57 has rank $\dfrac{1+2+3}{3} = 2$, 58 has rank 4, and 61 has rank 5.

3.   The efficiency rating of 0.91 indicates that with all other factors being the same, rank correlation requires 100 pairs of sample observations to achieve the same results as 91 pairs of observations with the parametric test for linear correlation, assuming that the stricter requirements for using linear correlation are met.

4.   The Wilcoxon rank-sum test does not require that the samples be from populations having a normal distribution or any other specific distribution.

5. $G = 4$

6.   Because there are only two runs, all of the values below the mean occur at the beginning and all of the values above the mean occur at the end, or vice versa. This indicates an upward (or downward) trend.

7.   Sign test and Wilcoxon signed-ranks test

8.   Rank correlation

9.   Kruskal-Wallis test

10.   Test claims involving matched pairs of data; test claims involving nominal data; test claims about the median of a single population

Review Exercises

1.   Using Excel's BINOM.DIST function, the two-tailed P-value $=2*\textbf{BINOM.DIST(44,106,0.5,TRUE)} = 0.0982$. The test statistic of $z = \dfrac{(44+0.5) - \frac{106}{2}}{\sqrt{106}/2} = -1.65$ is not less than or equal to the critical value of $z = -1.96$. Fail to reject the null hypothesis of $p = 0.5$. There is not sufficient evidence to warrant rejection of the claim that in each World Series, the American League team has a 0.5 probability of winning.

2.   Enter the data in an Excel worksheet. Select the **XLSTAT** add-in. Select **Nonparametrics tests**. Select **Comparison of two samples**. In the data ranges in the Sample boxes. Under Data format, check **Paired samples**. Select **Column labels**. Select **Sign test**. Click the **Options** tab. Alternative hypothesis: **Sample 1 – Sample 2 $\neq$ D**. Hypothesized difference (D): **0**. Significance level (%): **5**. Select **Exact p-value**. Click **OK**. The two-tailed P-value $= 0.0156$. There are 7 positive signs, 0 negative signs, 0 ties, and $n = 7$. The test statistic of $x = 0$ is less than or equal to the critical value of 0. There is sufficient evidence to reject the claim of no difference. It appears that there is a difference in cost between flights scheduled 1 day in advance and those

scheduled 30 days in advance. Because all of the flights scheduled 30 days in advance cost less than those scheduled 1 day in advance, it is wise to schedule flights 30 days in advance.

Sign test / Two-tailed test:	
N+	7
Expected value	3.5000
Variance (N+)	1.7500
p-value (Two-tailed)	0.0156
alpha	0.05

3. Enter the data in an Excel worksheet. Select the **XLSTAT** add-in. Select **Nonparametrics tests**. Select **Comparison of two samples**. In the data ranges in the Sample boxes. Under Data format, check **Paired samples**. Select **Column labels**. Select **Wilcoxon signed-rank test**. Click the **Options** tab. Alternative hypothesis: **Sample 1 – Sample 2 ≠ D**. Hypothesized difference (D): **0**. Significance level (%): **5**. Select **Exact p-value**. Click **OK**. The two-tailed P-value = 0.0156. There is sufficient evidence to reject the claim that differences between fares for flights scheduled 1 day in advance and those scheduled 30 days in advance have a median equal to 0. Because all of the flights scheduled 1 day in advance have higher fares than those scheduled 30 days in advance, it appears that it is generally less expensive to schedule flights 30 days in advance instead of 1 day in advance.

Wilcoxon signed-rank test / Two-tailed test:			
V	28		
Expected value	14.0000		
Variance (V)	35.0000		
p-value (Two-tailed)	0.0156		
alpha	0.05		

4. The sample mean is 54.8 years. $n_1 = 19$, $n_2 = 19$, and the number of runs is $G=18$. The critical values are 13 and 27 (From Table A-10.) Fail to reject the null hypothesis of randomness. There is not sufficient evidence to warrant rejection of the claim that the sequence of ages is random relative to values above and below the mean. The results do not suggest that there is an upward trend or a downward trend.

5. Enter the data in an Excel worksheet. Select the **XLSTAT** add-in. Select **Correlation/Association tests**. Select **Correlation tests**. In the **Observations/variables table** box, enter the data range. Type of Correlation: **Spearman**. Click **Variable labels**. Significance level (%): **5**. Click **OK**. Fail to reject the null hypothesis of $\rho_s = 0$. There is not sufficient evidence to support the claim that there is a correlation between the student ranks and the magazine ranks. When ranking colleges, students and the magazine do not appear to agree.

Correlation matrix (Spearman):		
Variables	Student	U.S.News & World Reports
Student	1	0.7143
U.S.News & World Reports	0.7143	1
p-values:		
Variables	Student	U.S.News & World Reports
Student	0	0.0576
U.S.News & World Reports	0.0576	0

6. Using Excel's BINOM.DIST function, the two-tailed P-value =**2*BINOM.DIST(13,32,0.5,TRUE)** = 0.3771. The test statistic of $z = \dfrac{(13 + 0.5) - \frac{32}{2}}{\sqrt{32}/2} = -0.88$ is not in the critical region bounded by $z = \pm 1.96$. There is not sufficient evidence to warrant rejection of the claim that the population of differences has a median of zero. Based on the sample data, it appears that the predictions are reasonably accurate, because there does not appear to be a difference between the actual high temperatures and the predicted high temperatures.

7. Enter the data in an Excel worksheet. Enter a column of 36 values all equal to **0**. Select the **XLSTAT** add-in. Select **Nonparametrics tests**. Select **Comparison of two samples**. Enter the data ranges in the Sample boxes. Under Data format, check **Paired samples**. If you included column labels in the data ranges, select **Column labels**. Select **Wilcoxon signed-ranks test**. Click the **Options** tab. Alternative hypothesis: **Sample 1 – Sample 2 $\neq$ D**. Hypothesized difference (D): **0**. Significance level (%): **5**. Select **Exact p-value**. Click **OK**. The two-tailed P-value = 0.5363. There is not sufficient evidence to warrant rejection of the claim that the population of differences has a median of zero. Based on the sample data, it appears that the predictions are reasonably accurate, because there does not appear to be a difference between the actual high temperatures and the predicted high temperatures.

Wilcoxon signed-rank test / Two-tailed test:		
V	230.5000	
Expected value	264.0000	
Variance (V)	2847.3750	
p-value (Two-tailed)	0.5363	
alpha	0.05	

8. Enter the data in an Excel worksheet. Select the **XLSTAT** add-in. Select **Nonparametrics tests**. Select **Comparison of K samples**. Enter data range in the **Samples** box. Under Data format, check **One column per sample**. Select **Column labels**. Select **Kruskal-Wallis test**. Click the **Options** tab. Significance level (%): **5**. Select **Asymptotic p-value**. Click **OK**. The P-value = 0.0351. Reject the null hypothesis of equal medians. Interbreeding of cultures is suggested by the data.

Kruskal-Wallis test:	
K (Observed value)	6.6980
K (Critical value)	5.9915
DF	2
p-value (Two-tailed)	0.0351
alpha	0.05

9. Enter the data in an Excel worksheet. Select the **XLSTAT** add-in. Select **Nonparametrics tests**. Select **Comparison of two samples**. Enter the data ranges in the Sample boxes. Under Data format, check **One column per sample**. Select **Column labels**. Select **Mann-Whitney test**. Click the **Options** tab. Alternative hypothesis: **Sample 1 – Sample 2 $\neq$ D**. Hypothesized difference (D): **0**. Significance level (%): **5**. Select **Asymptotic p-value**. Click **OK**. The two-tailed P-value = 0.0268. Reject the null hypothesis that the populations have the same median. Skull breadths from 4000 b.c. appear to have a different median than those from a.d. 150.

Mann-Whitney test / Two-tailed test:		
U	15.0000	
Expected value	40.5000	
Variance (U)	127.4559	
p-value (Two-tailed)	0.0268	
alpha	0.05	

10. Enter the data in an Excel worksheet. Select the **XLSTAT** add-in. Select **Correlation/Association tests**. Select **Correlation tests**. In the **Observations/variables table** box, enter the data range. Type of Correlation: **Spearman**. Click **Variable labels**. Significance level (%): **5**. Click **OK**. There is not sufficient evidence to support the claim that there is a correlation between weights of plastic and weights of food.

Correlation matrix (Spearman):

Variables	Plastic	Food
Plastic	1	0.4729
Food	0.4729	1

p-values:

Variables	Plastic	Food
Plastic	0	0.1206
Food	0.1206	0

Cumulative Review Exercises

1. Enter the credit hours data in an Excel worksheet. At the top of the screen, click **DATA**. Select **Data Analysis** in the Analysis group. Select **Descriptive Statistics**. Click **OK**. Click the **Input Range** box and enter the range of the credit hours data. If you included a label, click **Labels in First Row**. Click **Summary statistics**. Click **OK**. $\bar{x} = 14.6$ hours, median = 15.0 hours, $s = 1.7$ hours, $s^2 = 2.9$ hour^2, range = 6.0 hours

Credit Hours	
Mean	14.55
Standard Error	0.38027
Median	15
Mode	15
Standard Deviation	1.700619
Sample Variance	2.892105
Kurtosis	-0.37453
Skewness	-0.05939
Range	6
Minimum	12
Maximum	18
Sum	291
Count	20

2. a. Convenience sample

   b. Because the sample is from one class of statistics students, it is not likely to be representative of the population of all fulltime college students.

   c. Discrete

   d. Ratio

3. H_0: $\mu = 14$ hours . H_1: $\mu > 14$ hours . Select the **XLSTAT** add-in. Select **Parametric tests**. Select **One-sample *t*-test and *z*-test**. Enter the credit hours data range in the Data box. Under Data format, select **One sample**. If you included a label, click **Column labels**. Select **Student's *t* test**. Click the **Options** tab. Alternative hypothesis: **Mean 1 > Theoretical mean**. Theoretical mean: **14**. Significance level (%): **5**. Click **OK**. P-value = 0.0822. Fail to reject H_0. There is not sufficient evidence to support the claim that the mean is greater than 14 hours.

Difference	0.5500
t (Observed value)	1.4463
t (Critical value)	1.7291
DF	19
p-value (one-tailed)	0.0822
alpha	0.05

4. Enter the credit hours data in an Excel worksheet. In another column of the worksheet, enter a column of 20 values all equal to 14. Select the **XLSTAT** add-in. Select **Nonparametrics tests**. Select **Comparison of two**

samples. In the **Sample 1** box, enter the data range for the credit hours data. In the **Sample 2** box, enter the data range of the column of 20 values all equal to 14. Under Data format, check **Paired samples**. If you included labels in the data ranges, select **Column labels**. Select **Sign test**. Click the **Options** tab. Alternative hypothesis: **Sample 1 – Sample 2 > D**. Hypothesized difference (D): **0**. Significance level (%): **5**. Select **Exact p-value**. Click **OK**. The one-tailed P-value = 0.0717. There is not sufficient evidence to support the claim that the sample is from a population with a median greater than 14 hours.

N+	12
Expected value	8.5000
Variance (N+)	4.2500
p-value (one-tailed)	0.0717
alpha	0.05

5. Select the **XLSTAT** add-in. Select **Parametric tests**. Select **One-sample *t*-test and *z*-test**. Enter the credit hours data range in the Data box. Under Data format, select **One sample**. If you included a label, click **Column labels**. Select **Student's *t* test**. Click the **Options** tab. Alternative hypothesis: **Mean 1 ≠ Theoretical mean**. Theoretical mean: **14**. Significance level (%): **5**. Click **OK**. The 95% CI: $13.8 \text{ hours} < \mu < 15.3 \text{ hours}$. We have 95% confidence that the limits of 13.8 hours and 15.3 hours contain the true value of the population mean.

95% confidence interval on the mean:		
(13.7541, 15.3459)		

6. Enter the data in an Excel worksheet. Select the **XLSTAT** add-in. Select **Correlation/Association tests**. Select **Correlation tests**. Enter the data range in the **Observations/variables table**. Type of correlation: **Pearson**. If you included labels, click **Variable labels**. Significance level (%): **5**. Click **OK**. There is not sufficient evidence to support the claim of a linear correlation between price and quality score. It appears that you don't get better quality by paying more.

Correlation matrix (Pearson):

Variables	Price	Quality
Price	1	0.2050
Quality	0.2050	1

p-values:

Variables	Price	Quality
Price	0	0.6968
Quality	0.6968	0

7. Enter the data in an Excel worksheet. Select the **XLSTAT** add-in. Select **Correlation/Association tests**. Select **Correlation tests**. Enter the data range in the **Observations/variables table**. Type of correlation: **Spearman**. If you included labels, click **Variable labels**. Significance level (%): **5**. Click **OK**. There is not sufficient evidence to support the claim that there is a correlation between price and rank.

Correlation matrix (Spearman):

Variables	Price	Rank
Price	1	-0.5429
Rank	-0.5429	1

p-values:

Variables	Price	Rank
Price	0	0.2972
Rank	0.2972	0

8. Select the **XLSTAT** add-in. Select **Parametric tests**. Select **Tests for one proportion**. Frequency: **229**. Sample size: **740**: Test proportion **0.25**. Data format: **Frequency**. Select z **test**. Click the **Options** tab. Alternative hypothesis: **Proportion – Test proportion ≠ D**. Hypothesized difference (D): **0**. Significance level (%): **5**. Variance (confidence interval): **Sample**. Confidence interval: **Wald**. Click **OK**. The 95% CI: $0.276 < p < 0.343$. Because the value of 0.25 is not included in the range of values in the confidence interval, the result suggests that the percentage of all such telephones that are not functioning is different from 25%.

95% confidence interval on the proportion (Wald):			
(0.2762,	0.3428)		

9. $n = \dfrac{\left[z_{\alpha/2}\right]^2 (0.25)}{E^2} = \dfrac{1.645^2 (0.25)}{0.02^2} = 1692$ (Tech: 1691)

10. There must be an error, because the rates of 13.7% and 10.6% are not possible with samples of size 100.

Chapter 14

Statistical Process Control

Section 14-2, Basic Skills and Concepts

1. No. If we know that the manufacture of quarters is within statistical control, we know that the three out-of-control criteria are not violated, but we know nothing about whether the specification of 5.670 g is being met. It is possible to be within statistical control by manufacturing quarters with weights that are very far from the desired target of 5.670 g.

3. To use an $\bar{x}$ chart without an R chart is to ignore variation, and amounts of variation that are too large will result in too many defective goods or services, even though the mean might appear to be acceptable. To use an R chart without an $\bar{x}$ chart is to ignore the central tendency, so the goods or services might not vary much, but the process could be drifting so that daily process data do not vary much, but the daily means are steadily increasing or decreasing.

5. By following the instructions above, you will obtain output that includes all the values requested in Exercise 5.

Estimation:				
	Phase1			
Mean (estimate)	267.1143			

X-bar chart (control limits):		R chart (control limits):		
	Phase1			Phase1
UCL	290.1581	UCL		105.7466
CL	267.1143	CL		54.9600
LCL	244.0704	LCL		4.1734

$\bar{\bar{x}} = 267.11$ lb, and $\bar{R} = 54.96$ lb.

For the R chart: $LCL = 4.17$ lb and $UCL = 105.74$ lb.

For the $\bar{x}$ chart: $LCL = 244.07$ lb and $UCL = 290.16$ lb.

7. If you follow the instructions given right before Exercise 5, you obtain an R chart as part of the output generated by XLSTAT. The R chart does not violate any of the out-of-control criteria, so the variation of the process appears to be within statistical control

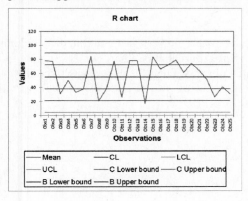

9. By following the instructions above, you will obtain output that includes all the values requested in Exercise 5.

Estimation:			
	Phase1		
Mean (estimate)	14.2496		
		R chart (control limits):	
X-bar chart (control limits):			
	Phase1		Phase1
UCL	14.3772	UCL	0.7356
CL	14.2496	CL	0.4140
LCL	14.1220	LCL	0.0924

$\bar{\bar{x}} = 14.250°C$, $\bar{R} = 0.414°C$.

For the R chart: $LCL = 0.092°C$ and $UCL = 0.736°C$.

For the $\bar{x}$ chart: $LCL = 14.122°C$ and $UCL = 14.377°C$.

11. If you follow the instructions given right before Exercise 9, you obtain an $\bar{x}$ chart as part of the output generated by XLSTAT. Because there is a pattern of an upward trend and there are points lying beyond the control limits, the $\bar{x}$ chart shows that the process is out of statistical control.

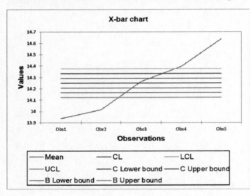

Section 14-2, Beyond the Basics

13. Enter the Table 14-1 data in an Excel worksheet. You do not need to enter the very first column, the $\bar{x}$ column, or the s column. Select the **XLSTAT** add-in. Select **SPC**. Select **Subgroup charts**. Chart family: **Subgroup charts**. Chart type: **S chart**. Click the **General** tab. Data format: **Columns**. Enter the data range in the **Data** box. Select **Sheet**. Select **Column labels**. Click the **Estimation** tab. Method for sigma (subgroups): S bar. Click **OK**. The R chart and the s chart are very similar in their pattern.

$\bar{s} = 0.0823$ g. $LCL = 0$ g and $UCL = 0.1719$ g .

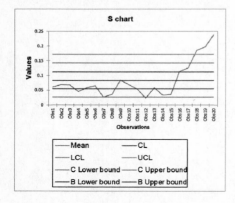

Section 14-3, Basic Skills and Concepts

1. No, the process does not appear to be within statistical control. There is a downward trend, there are at least 8 consecutive points all lying above the centerline, and there are at least 8 consecutive points all lying below the centerline. Because the proportions of defects are decreasing, the manufacturing process is not deteriorating; it is improving.

3. LCL denotes the lower control limit. Because the value of –0.000025 is negative and the actual proportion of defects cannot be less than 0, we should replace that value by 0.

5. Enter the data in an Excel worksheet. Select the **XLSTAT** add-in. Select **SPC**. Select **Attribute charts**. Chart family: **Attributes charts**. Chart type: **P chart**. Click the **General** tab. Enter the data range in the **Data** box. Common subgroup size: 10000. Click **Sheet**. If you included a label in the data range, click **Column labels**. Click **OK**. The process appears to be within statistical control.

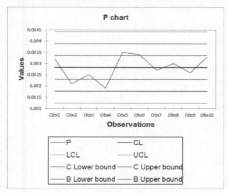

P chart (control limits):

	Phase1
UCL	0.0044
CL	0.0028
LCL	0.0012

Estimation:

	Phase1
Mean (estimate)	0.0028

7. Enter the data in an Excel worksheet. Select the **XLSTAT** add-in. Select **SPC**. Select **Attribute charts**. Chart family: **Attributes charts**. Chart type: **P chart**. Click the **General** tab. Enter the data range in the **Data** box. Common subgroup size: 100000. Click **Sheet**. If you included a label in the data range, click **Column labels**. Click **OK**. Because there appears to be a pattern of a downward shift and there are at least 8 consecutive points all lying above the centerline, the process is not within statistical control.

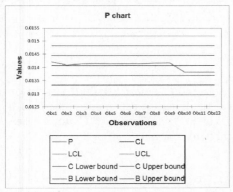

P chart (control limits):

	Phase1
UCL	0.0152
CL	0.0141
LCL	0.0130

Estimation:

	Phase1
Mean (estimate)	0.0141

9. Enter the data in an Excel worksheet. Select the **XLSTAT** add-in. Select **SPC**. Select **Attribute charts**. Chart family: **Attributes charts**. Chart type: **P chart**. Click the **General** tab. Enter the data range in the **Data** box. Common subgroup size: 1000. Click **Sheet**. If you included a label in the data range, click **Column labels**. Click **OK**. The process is out of control because there are points lying beyond the control limits and there are at least 8 points all lying below the centerline. The percentage of voters started to increase in recent years, and it should be much higher than any of the rates shown.

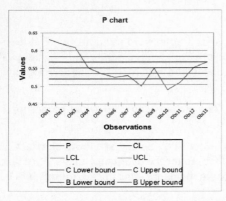

P chart (control limits):		
	Phase1	
UCL	0.5995	
CL	0.5523	
LCL	0.5051	

Estimation:	
	Phase1
Mean (estimate)	0.5523

11. Enter the data in an Excel worksheet. Select the **XLSTAT** add-in. Select **SPC**. Select **Attribute charts**. Chart family: **Attributes charts**. Chart type: **P chart**. Click the **General** tab. Enter the data range in the **Data** box. Common subgroup size: 500. Click **Sheet**. If you included a label in the data range, click **Column labels**. Click **OK**. There is a pattern of a downward trend and there are at least 8 consecutive points all below the centerline, so the process does not appear to be within statistical control. Because the rate of defects is decreasing, the process is actually improving and we should investigate the cause of that improvement so that it can be continued.

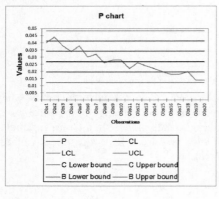

P chart (control limits):		
	Phase1	
UCL	0.0485	
CL	0.0268	
LCL	0.0051	

Estimation:	
	Phase1
Mean (estimate)	0.0268

Section 14-3, Beyond the Basics

13. Enter the data in an Excel worksheet. Select the **XLSTAT** add-in. Select **SPC**. Select **Attribute charts**. Chart family: **Attributes charts**. Chart type: **NP chart**. Click the **General** tab. Enter the data range in the **Data** box. Common subgroup size: 10000. Click **Sheet**. If you included a label in the data range, click **Column labels**. Click **OK**. Except for the vertical scale, the control chart is identical to the one obtained for Example 1.

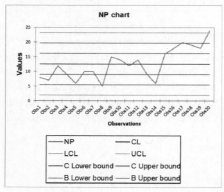

NP chart (control limits):		
	Phase1	
UCL	23.2422	
CL	12.6000	
LCL	1.9578	

Estimation:	
	Phase1
Mean (estimate)	12.6000

Chapter Quick Quiz

1. Process data are data arranged according to some time sequence. They are measurements of a characteristic of goods or services that result from some combination of equipment, people, materials, methods, and conditions.

2. Random variation is due to chance, but assignable variation results from causes that can be identified, such as defective machinery or untrained employees.

3. There is a pattern, trend, or cycle that is obviously not random. There is a point lying outside of the region between the upper and lower control limits. There are at least 8 consecutive points all above or all below the centerline.

4. An R chart uses ranges to monitor variation, but an $\bar{x}$ chart uses sample means to monitor the center (mean) of a process.

5. No. The R chart has at least 8 consecutive points all lying below the centerline and there are points lying beyond the upper control limit. Also, there is a pattern showing that the ranges have jumped in value for the most recent samples.

6. $\bar{R} = 52.8$ ft. In general, a value of $\bar{R}$ is found by first finding the range for the values within each individual subgroup; the mean of those ranges is the value of $\bar{R}$.

7. No. The $\bar{x}$ chart has a point lying below the lower control limit.

8. $\bar{\bar{x}} = 3.95$ ft. In general, a value of $\bar{\bar{x}}$ is found by first finding the mean of the values within each individual subgroup; the mean of those subgroup means is the value of $\bar{\bar{x}}$.

9. A p chart is a control chart of the proportions of some attribute, such as defective items.

10. Because there is a downward trend, the process is not within statistical control, but the rate of defects is decreasing, so we should investigate and identify the cause of that trend so that it can be continued.

Review Exercises

Follow these instructions to obtain the output you will need for Exercises 1 and 3. Select the **XLSTAT** add-in. Select **SPC**. Select **Subgroup charts**. Chart family: **Subgroup charts**. Chart type: **X bar – R chart**. Click the **General** tab. Data format: **Columns**. Enter the data range in the **Data** box. Select **Sheet**. If you included labels in the data range, select **Column labels**. Click the **Estimation** tab. Method for sigma (subgroups): R bar. Click **OK**.

1. By following the instructions above, you will obtain output that includes all the values requested in Exercise 1.

   $\bar{\bar{x}} = 2781.71$ kWh, $\bar{R} = 1729.38$ kWh.
   For the R chart: $LCL = 0$ kWh and $UCL = 3465.29$ kWh.
   For the $\bar{x}$ chart: $LCL = 1945.99$ kWh and $UCL = 3617.42$ kWh.

Estimation:		X-bar chart (control limits):		R chart (control limits):	
			Phase1		Phase1
		UCL	3617.4219	UCL	3465.2919
	Phase1	CL	2781.7083	CL	1729.3750
Mean (estimate)	2781.7083	LCL	1945.9947	LCL	0.0000

2. If you follow the instructions given right before Exercise 1, you obtain an R chart as part of the output generated by XLSTAT. The process variation is within statistical control.

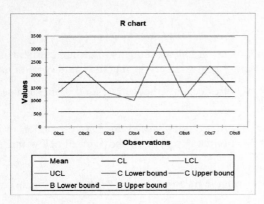

3. If you follow the instructions given right before Exercise 1, you obtain an $\bar{x}$ chart as part of the output generated by XLSTAT. The process mean is within statistical control.

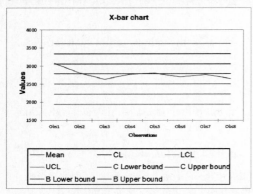

4. Rearrange the data so that it is placed in a single column starting with the Year 1 values and ending with the Year 8 values. Click and drag over the data range to select these data for the chart. At the top of the screen, click **INSERT**. Select **Line Chart** in the Chart group. Select the leftmost figure in the middle row. There does not appear to be a pattern suggesting that the process is not within statistical control. There is 1 point that appears to be exceptionally low. (The author's power company made an error in recording and reporting the energy consumption for that time period.)

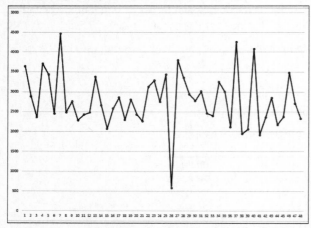

5. Enter the data in an Excel worksheet. Select the **XLSTAT** add-in. Select **SPC**. Select **Attribute charts**. Chart family: **Attributes charts**. Chart type: **P chart**. Click the **General** tab. Enter the data range in the **Data** box. Common subgroup size: 100. Click **Sheet**. If you included a label in the data range, click **Column labels**. Click **OK**. Because there are 8 consecutive points above the centerline and there is an upward trend, the process does not appear to be within statistical control.

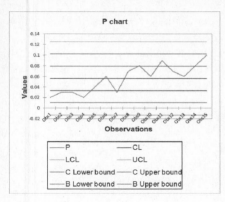

P chart (control limits):		
	Phase1	
UCL	0.1250	
CL	0.0560	
LCL	-0.0130	

Cumulative Review Exercises

1. Select the **XLSTAT** add-in. Select **Parametric tests**. Select **Tests for one proportion**. Frequency: **550**. Sample size: **1000**. Test proportion: **0.5**. Data format: **Frequency**. Select z **test**. Click the **Options** tab. Alternative hypothesis: **Proportion – Test proportion $\neq$ 0**. Hypothesized difference (D): **0**, Significance level (%): **5**. Variance (confidence interval): **Sample**. Confidence interval: **Wald**. Click **OK**. $0.519 < p < 0.581$. Because all of the values in the confidence interval estimate of the population proportion are greater than 0.5, it does appear that the majority of adults believe that it is not appropriate to wear shorts at work.

95% confidence interval on the proportion (Wald):			
(0.5192, 0.5808)			

2. a. $1 - 0.55 = 0.45$

    b. $(0.55)^5 = 0.0503$

    c. $1 - (0.55)^5 = 0.950$

3. Enter the data in an Excel worksheet. Be sure to include labels. Select the **XLSTAT** add-in. Select **Correlation/Association tests**. Select **Correlation tests**. Enter the data range in the **Observations/variables table**. Type of correlation: **Pearson**. Select **Variable labels**. Significance level (%): **5**. Click **OK**. There is sufficient evidence to support the claim that there is a linear correlation between yields from regular seed and kiln-dried seed. The purpose of the experiment was to determine whether there is a difference in yield from regular seed and kiln-dried seed (or whether kiln-dried seed produces a higher yield), but results from a test of correlation do not provide us with the information we need to address that issue.

Correlation matrix (Pearson):

Variables	Regular	Kiln-dried
Regular	1	0.8195
Kiln-dried	0.8195	1

p-values:

Variables	Regular	Kiln-dried
Regular	0	0.0020
Kiln-dried	0.0020	0

4. Enter the data in an Excel worksheet. Be sure to include labels. Select the **XLSTAT** add-in. Select **Parametric tests**. Select **Two-sample t-test and z-test**. Enter the range of the Regular data in the **Sample 1** box. Enter the range of the Kiln-dried data in the **Sample 2** box. Data format: **Paired samples**. Click **Column labels**. Select **Student's t test**. Click the **Options** tab. Alternative hypothesis: **Mean 1 – Mean 2 < 0**. Hypothesized difference

(D): **0**. Significance level (%): **5**. Click **OK**. H_0: $\mu_d = 0$. H_1: $\mu_d < 0$. Test statistic: $t = -1.532$. Critical value: $t = -1.8125$ (assuming a 0.05 significance level). P-value = 0.0783. Fail to reject H_0. There is not sufficient evidence to support the claim that kiln-dried seed is better in the sense that it produces a higher mean yield than regular seed. (The sign test can be used to arrive at the same conclusion; the test statistic is $x = 3$ and the critical value is 1. Also, the Wilcoxon signed-ranks test can be used; the test statistic is $T = 13.5$ and the critical value is 8.)

Difference	-1.0909
t (Observed value)	-1.5319
t (Critical value)	-1.8125
DF	10
p-value (one-tailed)	0.0783
alpha	0.05

5.   Enter the data in an Excel worksheet. Be sure to include labels. Select the **XLSTAT** add-in. Select **Describing data**. Select **Descriptive statistics**. Click **Quantitative data** and enter the range of the Regular and Kiln-dried data including the labels. Select **Sample labels**. Click the **Outputs** tab. Select **Mean** and **Standard deviation (n-1)**. Click **OK**. For the sample of yields from regular seed, $\bar{x} = 20.0$ and for the sample of yields from kiln-dried seed, $\bar{x} = 21.0$, so there does not appear to be a significant difference. For the sample of yields from regular seed, s = 3.4 and for the sample of yields from kiln-dried seed, s = 4.1, so there does not appear to be a significant difference.

Statistic	Regular	Kiln-dried
Mean	19.9545	21.0455
Standard deviation (n-1)	3.4493	4.1198

6.   Enter the data in an Excel worksheet. Select the **XLSTAT** add-in. Select **SPC**. Select **Attribute charts**. Chart family: **Attributes charts**. Chart type: **P chart**. Click the **General** tab. Enter the data range in the **Data** box. Common subgroup size: 50. Click **Sheet**. If you included a label in the data range, click **Column labels**. Click **OK**. There appears to be a pattern of an upward trend, so the process is not within statistical control.

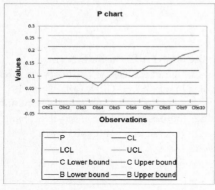

P chart (control limits):		
	Phase1	
UCL	0.2609	
CL	0.1220	
LCL	-0.0169	

7.   Enter the data in an Excel worksheet in order of highest to lowest percent. Click and drag over the data range to select the data for the chart. At the top of the screen, click **INSERT**. Select **Column chart** in the Charts group. Select the leftmost figure in the top row. You may want to use Excel's chart tools to add titles to the Pareto chart.

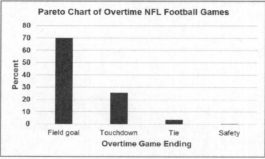

8. a.   Using Excel's NORM.DIST function, the percent of males with head breadths greater than 17 cm
         =1-NORM.DIST(17,15.2,2.5,TRUE) = 0.2358 or 23.58%. With 23.58% of males with head breadths
         greater than 17 cm, too many males would be excluded.

    b.   Using Excel's NORM.INV function, the 5th percentile **=NORM.INV(0.05,15.2,2.5)** = 11.08 cm.

         Using Excel's NORM.INV function, the 95th percentile **=NORM.INV(0.95,15,2.5)** = 19.31 cm.

9. With a voluntary response sample, the subjects decide themselves whether to be included. With a simple
    random sample, subjects are selected through some random process in such a way that all samples of the same
    size have the same chance of being selected. A simple random sample is generally better for use with statistical
    methods.

10. Sampling method (part c)